# MATHEMATICAL ASSET MANAGEMENT

# MATHEMATICAL ASSET MANAGEMENT

Thomas Höglund

WILEY-INTERSCIENCE
A JOHN WILEY & SONS, INC., PUBLICATION

Published by John Wiley & Sons, Inc., Hoboken, New Jersey
Published simultaneously in Canada



***Library of Congress Cataloging-in-Publication Data:***

Höglund, Thomas.
Mathematical asset management / Thomas Hoglund.
p. cm.
Includes bibliographical references and index.
ISBN 978-0-470-23287-3 (cloth)
1. Derivative securities--Mathematical models. 2. Risk management--Mathematical models. 3. Investment analysis--Mathematical models. I. Title.
HG6024.A3H63 2008
332.601'5195--dc22

2007050157

Printed in the United States of America

10 9 8 7 6 5 4 3 2 1

# CONTENTS

# PREFACE

This book is about risk management by means of trading strategies. It is intended for students in mathematics and finance, but I also hope that practitioners will find it useful.

I use the mathematics I think is relevant for our purposes — no more no less. Thus, for example, there is no Itô calculus and no martingales. On the other hand, there are a few results that require some mathematical maturity. As a consequence, the level of mathematics varies considerably, from elementary to fairly advanced.

The fact that the mathematics is rigorous does not imply that that the text lacks connection with the real world. On the contrary, every statement of some importance is checked against the reality.

The text is permeated by plots, tables, and numbers illustrating how the methods work in real life. The reader is strongly recommended to download the development of a few stocks and a relevant index. Then use these data to check how a trading strategy of her/his own choice works in practice (as done in Chapter 3), as well as to construct optimal portfolios (as done in Chapter 7).

In order to avoid burdening this book with too many details, there are many exercises scattered through the text. These are a vital part of the book. Solutions to the theoretical exercises, along with answers to the others, are given at the end of the book.

Since the year 2000 I have given two courses at Stockholm University on financial mathematics: one on derivatives and the other on portfolio selection. Both were based on this text, and I therefore know that the major part of the book can be understood by undergraduate students in mathematics.

The course on derivatives was based on Sections 1.1 and 1.2–1.2.3, Chapters 2 and 3, Sections 4.1–4.1.2, and the one-dimensional version of Model B in Section 4.3, and Chapter 5, whereas the course on portfolio choice was based on the remainder of the book together with Sections 1.1 and 4.1–4.1.2.

The material can be selected to suit different courses. For example, an elementary course on both derivatives and portfolio selection can be based on Sections 1.1 and 1.2–1.2.3 replacing the proof of the arbitrage theorem by a picture in two or three dimensions, Sections 2.1–2.3.2 with the omission of Proposition 2.3 and parts of Propositions 2.1 and 2.2, Sections 3.1–3.2, Sections 4.1–4.1.2 and 4.2, the one-dimensional version of Model B in Section 4.3, Sections 5.1–5.2.1 replacing the proof of Theorem 5.2 by a heuristic argument, Sections 6.1–6.1.1, 6.2–6.2.3, 7.1, 7.2.1, and 7.3–7.3.2 up to and including Exercise 7.12.

Adequate prerequisites for the major part of the book are calculus including Taylor's formula in several variables and Lagrange multipliers, elementary linear algebra, and probability theory including the central limit theorem.

THOMAS HÖGLUND

*Stockholm, Sweden*
*February 2008*

CHAPTER 1

# INTEREST RATE

Interest rate is the basis for financial transactions. Not only is there one rate at a given time, but the rate depends on the duration or time to maturity. This has interesting consequences as we shall see in Section 1.2. We shall in Section 1.1 start with the simple case when the interest rate is *flat*, which means that it is the same regardless its duration.

One essential thing in this chapter is to give an understanding of (a) the connection between present value and arbitrage and (b) how to construct arbitrage.

## 1.1 FLAT RATE

### 1.1.1 Compound Interest

If the interest is compounded $n$ times a year each time with the rate $r/n$, the value of one EUR after $t$ years equals

$$R = (1 + \frac{r}{n})^{nt} \quad \text{for } n = 1, 2, 3, ... \quad \text{and} \quad R = \lim_{n \to \infty} (1 + \frac{r}{n})^{nt} = e^{rt}$$

when the interest is compounded continuously.

For $r = 5\%$ and $t = 1$ these values are given by Table 1.1. These values thus depend

**Table 1.1.** The interest rate compounded $n$ times a year

| $n$ | Compounding every | | Value |
|---|---|---|---|
| 1 | year | $1.05$ | $= 1.050000$ |
| 2 | half year | $(1 + \frac{0.05}{2})^2$ | $= 1.050625$ |
| 4 | quarter | $(1 + \frac{0.05}{4})^4$ | $= 1.050945$ |
| 12 | month | $(1 + \frac{0.05}{12})^{12}$ | $= 1.051161$ |
| 52 | week | $(1 + \frac{0.05}{52})^{52}$ | $= 1.051245$ |
| 365 | day | $(1 + \frac{0.05}{365})^{365}$ | $= 1.051267$ |
| $\infty$ | continuously | $e^{0.05}$ | $= 1.051271$ |

not only on $r$ but also on $n$. The essential thing here is the *growth factor*, $R$. This can also be expressed by the interest rate, $r$, but then we must specify which interest rate we have in view. The most common is perhaps to define the rate as the *return*

$$r_1 = R - 1.$$

EXERCISE 1.1. Show that if the return equals $r'$ for one part of a period and $r''$ for the remainder, then the return for the entire period equals $r' + r'' + r'r''$.

The return is thus not additive, but this holds, however, for the *continuous rate*:

$$r_\infty = \ln R.$$

EXERCISE 1.2. Show that if the continuous rate equals $r'$ for one part of a period and $r''$ for the remainder, then the continuous rate for the entire period equals $r' + r''$.

We have $R = e^{r_\infty t}$. The continuous rate can therefore also be defined as the instantaneous return per time unit:

$$\lim_{t \to 0} \frac{e^{r_\infty t} - 1}{t} = r_\infty.$$

### 1.1.2 Present Value

$X_0$ EUR today is worth $X_T$ EUR in $T$ years. Here

$$X_T = R_T X_0$$

the *future value* of $X_0$ and

$$X_0 = d_T X_T$$

is the *present value* of $X_T$. Here $R_T$ is the growth factor during $T$ years and

$$d_T = R_T^{-1}$$

is the *discount factor*. We shall also write

$$X_0 = PV(X_T).$$

In order to value future payments, one compares their present values.

EXERCISE 1.3. Compare the value of 417 EUR in one year and 430 EUR in two years with 395 EUR today if the yearly return equals $5\%$.

EXERCISE 1.4. Express the time it takes to double a capital as a function of the continuous rate. In particular: How long will it take when the interest rate equals $5\%$? We shall sometimes also write $d(0, T)$ instead of $d_T$; and, in general,

$$d(t, T) = d_T / d_t$$

for the discount factor from time $t$ to time $T$.

### 1.1.3 Cash Streams

A *cash stream* is a sequence of real numbers, $\boldsymbol{x} = (x_0, x_1, ..., x_n)$, together with a sequence of points of time $0 = t_0 < t_1 < \cdots < t_n$. The holder of the cash stream receives $x_i$ EUR at time $t_i$. (This means that the holder pays $|x_i|$ EUR if $x_i < 0$.) The opposite party, the writer of the cash stream, holds the cash stream $-\boldsymbol{x}$.

The payments are thus divided into $n$ periods; $(t_{i-1}, t_i)$, $i = 1, ..., n$.

Here are three examples of cash streams:

**Loan.** You borrow today $S$ EUR and pay back $K$ EUR at the end of every period. This corresponds to the cash stream

$$(S, -K, ..., -K).$$

**Saving.** You put $K$ EUR in the bank at the beginning of every period, and you take out the entire amount at the end of the last period. This gives the cash stream

$$(-K, ..., -K, S).$$

**Annuity.** You put in $S$ EUR today and receive $K$ EUR at the end of each period. This gives the cash stream

$$(-S, K, ..., K).$$

This is also the cash stream of the lender when you take a loan.

Unless otherwise stated, we shall assume that the periods have the same lengths; $t_0 = 0, t_1 = 1, ..., t_n = n$ in some unit; day, month, or year for example. It is always possible to achieve this by letting $x_k = 0$ for certain $k$. The discount factor per period is in this case denoted by $d$. The present value of the cash stream is therefore given by

$$PV(\boldsymbol{x}) = x_0 + dx_1 + \cdots + d^n x_n.$$

EXERCISE 1.5. You receive 2000 EUR every year for 10 years with the first payment in one year. Calculate the present value of this cash stream if the rate of return is $5\%$ per year.

EXERCISE 1.6. A certain crop will give $1.05$ EUR back for each invested EUR when harvested after one year. When the crop is harvested after two or three years, the corresponding figures are $1.11$ and $1.14$, respectively. Compare these cash streams, assuming that the entire income will be reinvested in planting.

### 1.1.4 Effective Rate

The *effective rate* is the rate for which the cash stream has the present value 0; it is therefore determined by the discount factor for which $PV(\boldsymbol{x}) = 0$, provided that this is uniquely determined.

EXERCISE 1.7. Show that if $x_0 > 0$ and $x_i < 0$ for $i = 1, ..., n$ (or if $x_i < 0$ for $i = 0, ..., n-1$ and $x_n > 0$), then the discount factor is uniquely determined. Also show that the rate is positive ($d < 1$) in these cases if and only if

$$|x_0| < \sum_{k=1}^{n} |x_k|, \text{ (or } |x_n| > \sum_{k=0}^{n-1} |x_k|).$$

Let $m$ denote the number of periods per year. The yearly discount factor then equals $d^m$ and the continuous rate therefore equals

$$m \ln \frac{1}{d}$$

per year, whereas the yearly return equals

$$\frac{1}{d^m} - 1.$$

EXERCISE 1.8. You borrow 1000 EUR and repay the loan by paying 507 EUR after one and two months. Determine the effective rate.

EXERCISE 1.9. Determine the effective rates for the cash streams in Exercise 1.6.

EXERCISE 1.10. Show that the effective rate for the loan and the saving above are given by discount factors that satisfies

$$d\frac{1-d^n}{1-d} = \frac{S}{K} \quad \text{respectively} \quad d^{-n}\frac{1-d^n}{1-d} = \frac{S}{K}.$$

In order to solve $d$ from equations of this type, one can use Newton's method to find zeros of a differentiable function, $F(x)$: Start with a number $x_0$ which you think is close to the zero. Calculate then $x_1, x_2, ...$ via the formula

$$x_k = x_{k-1} - \frac{F(x_{k-1})}{F'(x_{k-1})},$$

for $k = 1, 2, ....$ This sequence converges toward a zero of $F$. For each iteration the number of correct decimals will double.

EXERCISE 1.11. Show that $x_k$ is the point in which the tangent of $F$ in the point $x_{k-1}$ cuts the $x$-axis, and use this to illustrate the construction of $x_1, x_2, ...$ graphically.

EXERCISE 1.12. You borrow 1000 EUR and repay by paying 338 EUR per month in three months. What is the effective rate?

EXERCISE 1.13. You borrow 20000 EUR in a bank and pay 300 EUR at the end of every month. The effective rate is given by 0.5% return per month. What is the yearly return? Determine the time it takes to pay back the loan. How much do you have to pay per month in order to repay the entire loan in 5 years?

### 1.1.5 Bonds

A *bond* is a cash stream of the form

$$(-P, c/m, ..., c/m, c/m + F).$$

Here the period length is $1/m$ years. The holder of the bond thus receives $c/m$ EUR $m$ times a year in $T = n/m$ years. $T$ is the *time to maturity*, $c$ the *coupon*, $F$ the *face value*, and $P$ the *price*. A *zero-coupon bond* is a bond with $c = 0$. A *T-bill* is a "zero-coupon bond" with maturity in at most a year.

The effective rate per year therefore is determined by the discount factor $d^m$, where $d$ satisfies

$$P = \frac{c}{m}\sum_{k=1}^{n} d^k + d^n F.$$

It is seen from this expression that the price is a decreasing function of the interest rate. Hence the prices of bonds decreases when the rate increases.

EXERCISE 1.14.

(a) Show that

$$P = \frac{c}{m} d \frac{1-d^n}{1-d} + d^n F.$$

(b) Define $y$ by

$$d = \frac{1}{1+\frac{y}{m}}.$$

That is, $y$ is the return during a period of length $1/m$ multiplied by $m$. Show that

$$P = \frac{c}{y} + d^n(F - \frac{c}{y}).$$

The number $y$ is called the *yield* of the bond. The above expression becomes especially simple when the yield is *par*, $c = yF$; $P = F$.

EXERCISE 1.15. Let $P_1$ and $P_2$ denote the prices of two bonds where the second has longer time to maturity than the first, but otherwise the two bonds are equal (the same coupon, yield, face value , and period). Show that

$$P_1 < P_2 \text{ for } y < c/F \quad \text{and} \quad P_1 > P_2 \text{ for } y > c/F.$$

EXERCISE 1.16. Calculate the effective rate for a five-year bond with face value 100 EUR, and yearly coupon payment 4 EUR, There are four coupon payments per year, and the price of the bond is 100 EUR.

One can form new cash streams by composing bonds into a portfolio.

EXERCISE 1.17. Consider two bonds having the same maturity date, period length and face value. One has the coupon $c_1$ and the other $c_2$, $c_1 < c_2$. The prices are $P_1$ respectively $P_2$.

(a) Use these bonds to construct a bond that has the coupon $c$, but the same face value. What is the price of this bond?

(b) Determine the weighs of the bonds in the portfolio so that the result becomes a zero-coupon bond.

(c) For which values of $c$ have both bonds positive weights in the portfolio?

### 1.1.6 The Effective Rate as a Measure of Valuation

The effective rate is a blunt tool to use to value cash streams in general. Consider the cash stream $\boldsymbol{x} = (ab, -a - b, 1)$.

The present value of this is

$$PV = ab - d(a + b) + d^2 = (d - a)(d - b).$$

This present value equals zero for $d = a$ and $d = b$. The effective rate is therefore not uniquely determined when $a \neq b$. Furthermore, the present value of the cash stream $-\boldsymbol{x}$ has the same zeros. It is therefore not clear how to use the effective rate to determine which of the two cash streams $\boldsymbol{x}$ and $-\boldsymbol{x}$ that is preferable (if any).

Assume that $a = 1$ and $b = 3$: $\boldsymbol{x} = (3, -4, 1)$. In this case the present value equals zero when $d = 1$ or $d = 3$. In the former case the effective rate equals zero, in the latter it is negative. The present value is positive for $\boldsymbol{x}$ and negative for $-\boldsymbol{x}$ when $d < 1$, which holds for normal interest rates. Therefore the cash stream $\boldsymbol{x}$ ought to be preferable to $-\boldsymbol{x}$.

Assume that the time between the payments is one year, that you can borrow money at the (return)rate $5\%$ per year, and that you can lend at the rate $4\%$. The following procedure shows that it is advantageous to hold $\boldsymbol{x}$:

> At $t = 0$: Accept the cash stream $\boldsymbol{x}$. You receive 3 EUR which you lend for one year at the rate $4\%$.
>
> At $t = 1$:. The loan is repaid, and you get $3 \times 1.04 = 3.12$ EUR. You borrow $0.88$ EUR for one year and pay 4 EUR.
>
> At $t = 2$: You receive 1 EUR and repay the loan with $0.88 \times 1.05 = 0.924$ EUR. Remains $0.076$ EUR.

In this way you get the cash stream $(0, 0, 0.076)$, and you have made a profit without taking any risk. That is, you have made *arbitrage*.

EXERCISE 1.18. In which way shall the lending rate be related to the interest on deposit in the above example in order to make it possible to make arbitrage in this way?

EXERCISE 1.19. You are offered two cash streams: $(1000, -3000, 2000)$ and $(-1000, 3000, -2000)$. The time between the payments is one year.

(a) Determine the effective rates of the cash streams.

(b) You can borrow at $5\%$ per year and lend at $4\%$. Describe how you can make arbitrage.

## 1.2 DEPENDENCE ON THE MATURITY DATE

We shall now take into account that the interest rate varies with the maturity date, and we shall investigate the consequences of this fact.

### 1.2.1 Zero-Coupon Bonds

The holder of a $k$-year zero-coupon bond receives 1 EUR after $k$ years. Let $d_k$ denote the price of the $k$-year zero-coupon bond, $k = 1, 2, ..., n$, and put $d_0 = 1$. The price $d_k$ defines the value today (the present value) of 1 EUR in $k$ years. These prices also define the $k$-year *spot rate*, $r_k$, by the relation

$$d_k = e^{-kr_k},$$

provided that we use the continuous rate. If instead $r_k$ stands for the yearly returns, we have the relation

$$d_k = (1 + r_k)^{-k}.$$

Here we shall simplify the reasoning and assume that we can buy and sell these bonds in arbitrary quantities without transaction costs. Thus, for example, if we want to sell a fraction of a bond we don't have, we can borrow this fraction without cost and sell it in order to buy it back later for a price that hopefully is lower (short-selling or shorting).

### 1.2.2 Arbitrage-Free Cash Streams

The present value of the cash stream $\boldsymbol{x} = (x_0, x_1, ..., x_n)$ is

$$PV(\boldsymbol{x}) = \sum_{k=0}^{n} d_k x_k.$$

Let $\boldsymbol{x}'$ denote the cash stream one gets by proceeding in the following way: Buy at $t = 0$ $x_k$ $k$-year bonds (this means that one sells $-x_k$ if $x_k < 0$), for $k = 1, ..., n$.

Wait until $t = n$. The cost at $t = 0$ is

$$\sum_{k=1}^{n} x_k d_k = PV(\boldsymbol{x}) - x_0$$

and one receives $x_k$ EUR at $t = k$, $k = 1, 2, ..., n$. Therefore

$$\boldsymbol{x}' = \boldsymbol{x} - \boldsymbol{p}, \text{ where } \boldsymbol{p} = (PV(\boldsymbol{x}), 0, ..., 0).$$

Assume that $PV(\boldsymbol{x}) > 0$. By accepting the cash streams $\boldsymbol{x}$ and $-\boldsymbol{x}'$, one gets the cash stream $\boldsymbol{p} = \boldsymbol{x} - \boldsymbol{x}'$ and hence a risk-free profit. If instead $PV(\boldsymbol{x}) < 0$, one gets the cash stream $-\boldsymbol{p}$ by accepting $-\boldsymbol{x}$ and $\boldsymbol{x}'$ and hence a risk-free profit even in this case.

This is called arbitrage. If we assume that it is not possible to make arbitrage, we thus must have $PV(\boldsymbol{x}) = 0$ for all cash streams. Observe that a geometrical way to express this is to say that $\boldsymbol{x}$ is orthogonal to the discount vector $\boldsymbol{d} = (d_0, d_1, ..., d_n)$; $\boldsymbol{x} \cdot \boldsymbol{d} = 0$.

EXERCISE 1.20. You intend to borrow 1000 EUR and have to choose between the following two alternatives: $\boldsymbol{x} = (1000, -866, -181)$ and $\boldsymbol{y} = (1000, -426, -656)$. The first repayment is in one year, and the second one is in two years. The one-year zero-coupon bond costs $0.97$ EUR, and the two-year bond costs $0.89$ EUR.

(a) Calculate the effective rates of the two loans.

(b) Calculate the present values of the two cash streams. Also calculate the one- and two-year spot rates.

(c) The loan $\boldsymbol{y}$ thus is preferable despite the fact that it has higher effective rate than $\boldsymbol{x}$. Describe how to make arbitrage by accepting $\boldsymbol{y}$.

(d) Also describe how the lender can make arbitrage if you accept the loan $\boldsymbol{x}$.

### 1.2.3 The Arbitrage Theorem

Forget for a moment the concrete interpretation of $\boldsymbol{d}$ as prices of bonds. We shall here instead show that there is a unique discount vector in each market that satisfies certain conditions. Let $\boldsymbol{x}_1$, $\boldsymbol{x}_2$,..., $\boldsymbol{x}_N$ be given cash streams in $\mathbb{R}^{n+1}$, and let $L$ denote the vector space generated by these. We shall say that there are *arbitrage* opportunities if there is a $\boldsymbol{x}$ in $L$ such that $\boldsymbol{x} \neq 0$ and $\boldsymbol{x} \geq 0$ (the latter means that $x_k \geq 0$ for all $k = 0, 1, ..., n$). That $L$ has no arbitrage opportunities can therefore be expressed in the following way:

$$L \cap \mathbb{R}_+^{n+1} = \{0\}.$$

In this case we must have $\dim(L) \leq n$. We shall also say that the market (i.e., $L$) is *complete* if $\dim(L) = n$.

**Theorem 1.1.** *The market L is complete and has no arbitrage opportunities if and only if there is a* $\mathbf{d}$ *in* $\mathbb{R}^{n+1}$ *with* $d_0 = 1$ *and* $d_1 > 0, d_2 > 0, ..., d_n > 0$ *such that* $L = \{\boldsymbol{x}; \boldsymbol{x} \cdot \mathbf{d} = 0\}$. *The discount vector* $\mathbf{d}$ *is uniquely determined by L.*

*Proof.* The theorem is geometrically obvious when $n = 1$ and 2. Isn't it?

Assume that $L$ is complete and has no arbitrage opportunities. Let $\mathbf{c} \neq 0$ be in the orthogonal complement of $L$; that is, $\mathbf{c} \cdot \boldsymbol{x} = 0$ for all $\boldsymbol{x}$ in $L$. Then $\mathbf{c}$ is uniquely determined up to a multiplicative constant and $L = \{\boldsymbol{x}; \mathbf{c} \cdot \boldsymbol{x} = 0\}$ since $L$ is complete. If $c_k = 0$, then $\boldsymbol{e}_k \cdot \mathbf{c} = 0$. Hence $\boldsymbol{e}_k \in L$, a contradiction. Hence $c_k \neq 0$ for all $k$. Assume that not all $c_k$ have the same sign. Then there are indices $i$ and $j$ such that $c_i > 0$ and $c_j < 0$. Put $\boldsymbol{u} = c_i \boldsymbol{e}_j - c_j \boldsymbol{e}_i$. Then $\boldsymbol{u} \neq 0$, $\boldsymbol{u} \geq 0$, and $\boldsymbol{u} \cdot \mathbf{c} = 0$; that is $\boldsymbol{u} \in L$, a contradiction. It follows that all $c_k$ have the same sign.

The only if part now follows with $d_k = c_k / c_0$. Conversely, it is clear that if $\boldsymbol{x} \cdot \boldsymbol{d} = 0$, then it is impossible that $\boldsymbol{x} \neq 0$ and $\boldsymbol{x} \geq 0$. □

Assume that $L$ is complete and has no arbitrage opportunities and that a new cash stream of the form $(-p, a_1, ..., a_n)$, where $a_1, ..., a_n$ are given numbers, is introduced on the market. This extended market is thus free of arbitrage if and only if $p = d_1 a_1 + \cdots + d_n a_n$.

The zero-coupon bonds $\boldsymbol{b}_1 = (-d_1, 1, 0, 0, ..., 0, 0)$, $\boldsymbol{b}_2 = (-d_2, 0, 1, 0, ...,, 0, 0)$, ..., $\boldsymbol{b}_n = (-d_n, 0, 0, 0, ..., 0, 1)$ is a basis for $L$. Aren't they?

EXERCISE 1.21. The market $L_1$ is generated by the two vectors $(5, -6, -6)$ and $(5, -5, -4)$, $L_2$ is generated by $(5, -6, -6)$ and $(-1, 0, 3)$, and $L_3$ is generated by $(5, -4, -2)$ and $(-8, 6, 3)$.

(a) Decide which of these markets that have no arbitrage opportunities.

(b) An additional cash stream of the form $(-p, 2, 3)$ is introduced on the market $L_2$. Determine its price (i.e., determine $p$) such that the extended market doesn't have any arbitrage opportunities.

EXERCISE 1.22. The market $L$ is generated by the cash stream $(2, -2, -1)$ and is therefore not complete. Another cash stream of the form $(-p, 1, 1)$ is going to be introduced. For which values of p will the extended market be complete? Without arbitrage opportunities?

EXERCISE 1.23. Consider the following three bonds where the period length is one year: $A = (-P_A, 110, 0, 0))$, $B = (-P_B, 10, 110, 0)$, $C = (-P_C, 10, 10, 110)$.

(a) Determine the price, $P_C$, and the effective rate (yield) for the bond $C$ if the spot rates (measured by the yearly returns) are as follows: 1 year $= 7\%$, 2 year $= 9\%$, 3 year $= 11\%$.

(b) Determine the prices and the spot rates if $A$, $B$, and $C$ have the effective rates $8.5\%$, $9.0\%$, $11.5\%$, respectively.

### 1.2.4 The Movements of the Interest Rate Curve

If we plot the yield of government bonds as a function of the maturity date, we will get a curve called the *yield curve*. The *spot rate curve* is the spot rate as a function of the maturity date. The two curves are similar but not identical because the latter is based only on zero-coupon bonds whereas the former is also based on coupon bonds. The spot rate curve is the relevant one for our purposes, but the yield curve is easier to determine. These curves are not constant but varies with time.

Frye (1997) made a statistical study of the daily changes of the yield of ten American government securities during 1543 days from 1989 and 1995. The result was

$$\partial \boldsymbol{r} = \xi_1 \boldsymbol{a}_1 + \cdots + \xi_{10} \boldsymbol{a}_{10}.$$

Here $\partial \boldsymbol{r} = (\partial r_1, ..., \partial r_{10})$ stands for the daily change of rate. The vectors $\boldsymbol{a}_1, ..., \boldsymbol{a}_{10}$ are given in Table 1.2.

**Table 1.2.** Orthonormal basis vectors for the change of rates

| Maturity | 3 mo | 6 mo | 1 yr | 2 yr | 3 yr | 4 yr | 5 yr | 7 yr | 10 yr | 30 yr |
|---|---|---|---|---|---|---|---|---|---|---|
| $\boldsymbol{a}_1$ | 0.21 | 0.26 | 0.32 | 0.35 | 0.36 | 0.36 | 0.36 | 0.34 | 0.31 | 0.25 |
| $\boldsymbol{a}_2$ | −0.57 | −0.49 | −0.32 | −0.10 | 0.02 | 0.14 | 0.17 | 0.27 | 0.30 | 0.33 |
| $\boldsymbol{a}_3$ | 0.50 | 0.23 | −0.37 | −0.38 | −0.30 | −0.12 | −0.04 | 0.15 | 0.28 | 0.46 |
| $\boldsymbol{a}_4$ | 0.47 | −0.37 | −0.58 | 0.17 | 0.27 | 0.25 | 0.14 | 0.01 | −0.10 | −0.34 |
| $\boldsymbol{a}_5$ | −0.39 | 0.70 | −0.52 | 0.04 | 0.07 | 0.16 | 0.08 | 0.00 | −0.06 | −0.18 |
| $\boldsymbol{a}_6$ | −0.02 | 0.01 | −0.23 | 0.59 | 0.24 | −0.63 | −0.10 | −0.12 | 0.01 | 0.33 |
| $\boldsymbol{a}_7$ | 0.01 | −0.04 | −0.04 | 0.56 | −0.79 | 0.15 | 0.09 | 0.13 | 0.03 | −0.09 |
| $\boldsymbol{a}_8$ | 0.00 | −0.02 | −0.05 | 0.12 | 0.00 | 0.55 | −0.26 | −0.54 | −0.23 | 0.52 |
| $\boldsymbol{a}_9$ | 0.01 | −0.01 | 0.00 | −0.12 | −0.09 | −0.14 | 0.71 | 0.00 | −0.63 | 0.26 |
| $\boldsymbol{a}_{10}$ | 0.00 | 0.00 | 0.01 | −0.05 | −0.00 | −0.08 | 0.48 | −0.68 | 0.52 | −0.13 |

These are pairwise orthogonal and have the length 1. The stochastic variables $\xi_1, ..., \xi_{10}$ are *uncorrelated* and ordered after decreasing standard deviations, $\sigma_1 > \sigma_2 > \cdots > \sigma_{10}$. These are given in Table 1.3.

**Table 1.3.** The standard deviations in decreasing order

| $i$ | 1 | 2 | 3 | 4 | 5 | 6 | 7 | 8 | 9 | 10 |
|---|---|---|---|---|---|---|---|---|---|---|
| $\sigma_i$ | 17.49 | 6.05 | 3.10 | 2.17 | 1.97 | 1.69 | 1.27 | 1.24 | 0.80 | 0.79 |

The unit is basis points (bp), that is, $1/100\% = 0.0001$.

The expected value of $\partial \boldsymbol{r}$ is negligible compared to the fluctuations, and therefore the standard deviation is the relevant measure.

We therefore have

$$E|\partial \boldsymbol{r}|^2 \approx \sigma^2 = \sigma_1^2 + \cdots + \sigma_{10}^2 = 367.9.$$

Movements of the yield along $\boldsymbol{a}_1$ explains $17.49^2/367.9 = 83\%$ of the total variance, and movements along $\boldsymbol{a}_2$ explains $10\%$, together $93\%$. By adding $\boldsymbol{a}_3$, one reaches $96\%$. In Fig. 1.1 is a plot of $\boldsymbol{a}_1, \boldsymbol{a}_2$, and $\boldsymbol{a}_3$. The first corresponds roughly to a parallel

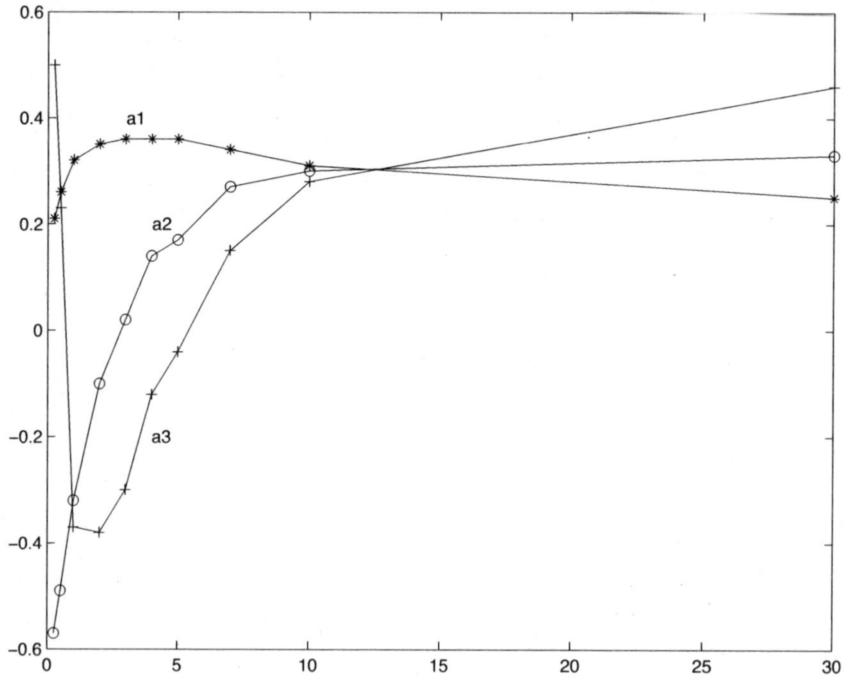


**Figure 1.1.** The three most important components that explains the movements of the yield curve.

shift in the yield curve, the second to steepening; rates with time to maturity shorter than approximately 2 years moves in one direction, and the other rates in the other direction. The third factor corresponds to a bend; short and long yields moves in one direction, and the other rates in the other direction.

We are not going to use the exact expressions for $\boldsymbol{a}_1, \boldsymbol{a}_2$, and $\boldsymbol{a}_3$, but we will approximate these with simple analytical expressions. Reasons for this is (except that is convenient):

1. The investigation comprises *one* market during *one* period, and it is not clear that one would get exactly the same result under other circumstances.

2. The investigation concern the yield curve and not the spot rate curve.
3. We shall consider the continuous rate rather than the return, and they are not exactly the same.

Assume that the rate changes from $\boldsymbol{r} = (r_1, ..., r_n)$ to $\boldsymbol{r} + \partial \boldsymbol{r}$. The analytical expressions that approximate $\partial \boldsymbol{r}$ can be chosen in different ways. Here we shall first consider *parallel shift*,

$$\partial \boldsymbol{r} = \mathbf{1}\partial p, \quad \text{where } \mathbf{1} = (1, ..., 1)$$

and then consider *steepening*,

$$\partial \boldsymbol{r} = \boldsymbol{r}\partial b.$$

By also considering changes of the form

$$\partial \boldsymbol{r} = \boldsymbol{r}^2 \partial c, \quad \text{where } \boldsymbol{r}^2 = (r_1^2, ..., r_n^2),$$

it is also possible to imitate a certain kind of *bending*.

By adding $\boldsymbol{r}^k = (r_1^k, ..., r_n^k)$ for $k = 3, 4, ...$ one can increase the precision (but also the complexity) in the approximation of $\partial \boldsymbol{r}$. At $k = n - 1$ the fit will be perfect.

EXERCISE 1.24. Assume that all rates, $r_1, ..., r_n$, are different. Show that $\boldsymbol{r}^k$, $k = 0, ..., n-1$ span $\mathbb{R}^n$.

It is clear that $\boldsymbol{a}_3$ cannot be imitated by a parabola, but this fact isn't crucial. We shall later use the results in this section to immunize bond portfolios. A possibility is to group the bonds by dividing the dates to maturity in a few intervals and treat each group separately.

In Fig. 1.2 a part of the yield curves for Swedish government securities are shown. The curves are from August 4 and September 4, 2000. We have adapted polynomial of degree 0, "o", degree 1, "*", and degree 2, "+" by least square to the lower curve.

The mean distance between the two yield curves,

$$d = \sqrt{\sum_1^5 (r_i' - r_i)^2/5},$$

is 31 basis points, whereas the mean distances between the lower curve and the different approximations are 7.6, 4.3, and 3.8, respectively. Parallel shift explains thus the major part of change of rate in this case.

### 1.2.5 Sensitivity to Change of Rates

The price of a $T$-year zero-coupon bond with face value 1 is given by

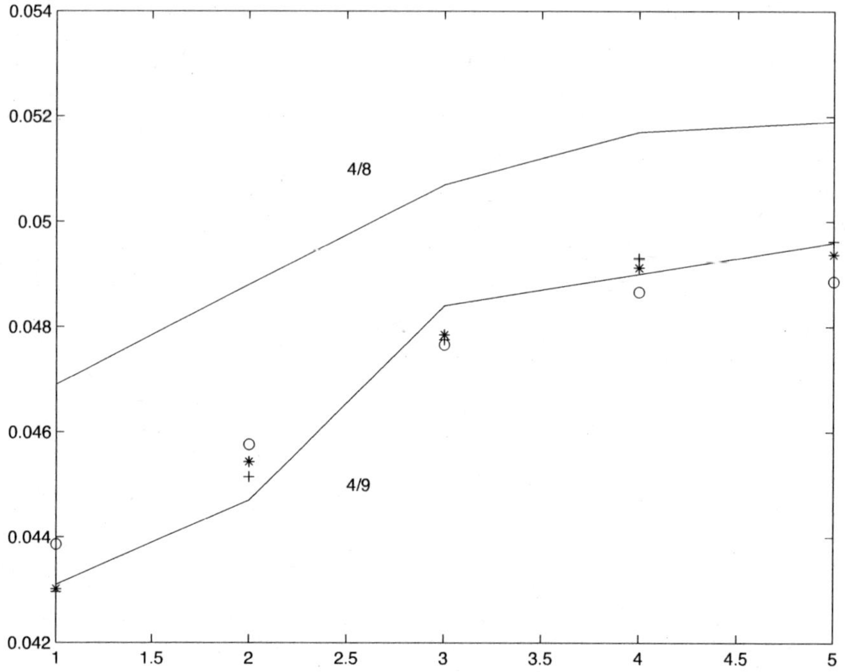


**Figure 1.2.** The change of the yield curve, and approximation with polynomials of degree 0, "o", degree 1, "*", and degree 2, "+".

$$P = e^{-rT},$$

where $r$ is the T-year spot rate. The change of the price of this bond when $r \to r + \partial r$ is given in the next exercise.

EXERCISE 1.25. Show that if one neglects terms of the order $(\partial r)^2$, then the following holds for the price of a zero-coupon bond with time to maturity $T$:

$$\frac{\partial P}{P} = -T\partial r$$

In this case the relative change of the price is proportional to the time to maturity. Zero-coupon bonds with long duration is thus especially sensitive to change of rates.

In order to study the effect of a change of rates $\boldsymbol{r} \to \boldsymbol{r} + \partial \boldsymbol{r}$, where $\boldsymbol{r} = (r_1, ..., r_n)$ and $\partial \boldsymbol{r} = (\partial r_1, ..., \partial r_n)$, on a general cash stream $\boldsymbol{x} = (-P, x_1, ..., x_n)$ we shall write

$$P(\boldsymbol{r}) = \sum_{k=1}^{n} d_k x_k$$

for the present value of the future payments as a function of $\boldsymbol{r}$. We have

$$P(\boldsymbol{r} + \partial\boldsymbol{r}) = P(\boldsymbol{r}) + \nabla P(\boldsymbol{r}) \cdot \partial\boldsymbol{r} + O\big(|\partial\boldsymbol{r}|^2\big).$$

EXERCISE 1.26. Show that

$$\frac{\partial P}{\partial r_k} = -k d_k x_k.$$

It follows that

$$\frac{\partial P}{P} = -\boldsymbol{D} \cdot \partial\boldsymbol{r} + O\big(|\partial\boldsymbol{r}|^2\big).$$

Here

$$\boldsymbol{D} = (1v_1, 2v_2, ..., nv_n) \quad \text{and} \quad v_k = \frac{d_k x_k}{P}.$$

In particular, if the spot rates change by a parallel shift, then

$$\frac{\partial P}{P} = -D\partial p + O\big(|\partial p|^2\big),$$

where

$$D = \boldsymbol{D} \cdot \mathbf{1} = 1v_1 + 2v_2 + \cdots + nv_n$$

is the *duration* of $\boldsymbol{x}$. Note that this is a weighted mean value of the payment time points $1, ..., n$, and where the weights are proportional to the present values of the payments. Also note that the duration of a zero-coupon bond with time to maturity $T$ equals $T$.

EXERCISE 1.27. Show that if two bond portfolios have the prices $P_1$ and $P_2$ and the durations $D_1$ and $D_2$, then the portfolio consisting of the two bonds has the duration

$$\frac{P_1}{P_1 + P_2} D_1 + \frac{P_2}{P_1 + P_2} D_2.$$

It follows that a bond portfolio, formed by a number of bonds each having positive weight, will have a duration between the smallest and largest of the durations of the individual bonds.

EXERCISE 1.28. Consider the following three bonds where the period length is one year: $A = (-P_A, 5, 5, 5, 5, 105)$, $B = (-P_B, 4, 4, 4, 104, 0)$, $C = (-P_C, 0, 100, 0, 0, 0)$. Calculate the prices and durations in the case where the (continuous) spot rates are $4.69\%$, $4.88\%$, $5.07\%$, $5.17\%$, and $5.19\%$.

EXERCISE 1.29. Consider a bond with face value $F$ and coupon $c$ which is paid $m$ times a year in $T = n/m$ years. The period length thus equals $1/m$ years. Assume that the rate is flat; $d_k = d^k$, where

$$d = e^{-r/m} = \frac{1}{1 + \frac{y}{m}}.$$

Show that

(a)

$$D = \frac{1}{m}\Big(\frac{c}{m}\sum_{k=1}^{n} kd^k + nd^n F\Big)/P \text{ years.}$$

(b)

$$\sum_{k=1}^{n} kd^{k-1} = \frac{1}{1-d}\Big(\frac{1-d^{n+1}}{1-d} - (n+1)d^n\Big)$$

(c)

$$D = \bigg(\frac{c}{y}\Big(\frac{1}{y} + \frac{1}{m}\Big)(1 - d^n) + T\Big(F - \frac{c}{y}\Big)d^n\bigg)/P$$

(d)

$$D = \Big(\frac{1}{y} + \frac{1}{m}\Big)(1 - d^n)$$

when $c = yF$.

If the spot rates change by steepening, $\partial \boldsymbol{r} = \boldsymbol{r}\partial b$, then

$$\frac{\partial P}{P} = -D^{(1)}\partial b + O\big(|\partial b|^2\big),$$

where

$$D^{(1)} = \boldsymbol{D} \cdot \boldsymbol{r} = r_1 v_1 + 2r_2 v_2 + \cdots + nr_n v_n$$

is a weighted mean value of the rates until the different payments.

In a corresponding way the change of rates $\partial \boldsymbol{r} = \boldsymbol{r}^2 \partial c$ will give the change of the price

$$\frac{\partial P}{P} = -D^{(2)}\partial b + O\big(|\partial c|^2\big),$$

where

$$D^{(2)} = \boldsymbol{D} \cdot \boldsymbol{r}^2 = r_1^2 v_1 + 2r_2^2 v_2 + \cdots + nr_n^2 v_n.$$

Change of rates of the form $\partial r_k = \pi\partial p + \beta r_k \partial b + \gamma r_k^2 \partial c$ will thus give the change of price

$$\frac{\partial P}{P} \approx -\pi D\partial p - \beta D^{(1)}\partial b - \gamma D^{(2)}\partial c.$$

EXERCISE 1.30. Calculate $D^{(1)}$ and $D^{(2)}$ for the bonds in Exercise 1.28.

Assume that Fig. 1.2 describes the spot rate (and not the yield). In Table 1.4 the relative change of price, $\partial P/P$, of the three bonds $A$, $B$, and $C$ are given. The values

$$-D\partial p, \ -D\partial p - D^{(1)}\partial b \text{ and } - D\partial p - D^{(1)}\partial b - D^{(2)}\partial c$$

are given in the three columns to the right. We have used the least-square estimates shown in the figure. The unit is basis points.

**Table 1.4.** The relative change of price

| | $\partial P/P$ | Parallel | +Steepening | +Bending |
|---|---|---|---|---|
| $A$ | 108 | 138 | 117 | 106 |
| $B$ | 103 | 115 | 99 | 92 |
| $C$ | 82 | 61 | 67 | 75 |

In this case the approximations are improved if one also immunizes against steepening, but only for $A$ and $C$ if one, in addition, immunizes against bending.

### 1.2.6 Immunization

Assume that we already today want to guarantee future payment obligations; $x_1$, $x_2, ..., x_n$, where $x_k$ shall be paid at time $k$. This can also be formulated in the following way: We hold the cash stream

$$\boldsymbol{x}_0 = (0, -x_1, ..., -x_n)$$

and want to replace this with a cash stream of the form

$$(-P, 0, ..., 0).$$

An imaginable possibility is that today buy $x_k$ zero-coupon bonds with face values 1 and time to maturity $k$, for $k = 1, ..., n$, that is, to provide the cash stream

$$\boldsymbol{x} = (-P, x_1, ..., x_n),$$

where

$$P = d_1 x_1 + ... + d_n x_n$$

is the price we have to pay for the bonds. Our payment obligations now have been reduced to

$$\boldsymbol{x}_0 + \boldsymbol{x} = (-P, 0, ..., 0),$$

and we have thus eliminated our future payment obligations, provided that the bonds are risk-free.

It may happen that there are no zero-coupon bonds with exactly these dates of maturity, or the payments may be so many and small that the above procedure is inconvenient.

An alternative in this case is to construct a bond portfolio, $\boldsymbol{y}$, that consists of fewer bonds and that may have different dates of maturity, but they are chosen in such a way that $\boldsymbol{y}$ has the same present value as $\boldsymbol{x}$ and reacts in a similar way on changes of rates.

We have

$$P_{\boldsymbol{y}}(\boldsymbol{r}+\partial\boldsymbol{r})-P_{\boldsymbol{x}}(\boldsymbol{r}+\partial\boldsymbol{r}) = P_{\boldsymbol{y}}(\boldsymbol{r})-P_{\boldsymbol{x}}(\boldsymbol{r})+\left(\nabla P_{\boldsymbol{y}}(\boldsymbol{r})-\nabla P_{\boldsymbol{x}}(\boldsymbol{r})\right)\cdot\partial\boldsymbol{r}+O\left(|\partial\boldsymbol{r}|^2\right)$$

and therefore

$$P_{\boldsymbol{y}}(\boldsymbol{r}+\partial\boldsymbol{r})-P_{\boldsymbol{x}}(\boldsymbol{r}+\partial\boldsymbol{r}) = O\left(|\partial\boldsymbol{r}|^2\right)$$

if

$$P_{\boldsymbol{y}}(\boldsymbol{r}) = P_{\boldsymbol{x}}(\boldsymbol{r}) \quad \text{and} \quad \left(\nabla P_{\boldsymbol{y}}(\boldsymbol{r})-\nabla P_{\boldsymbol{x}}(\boldsymbol{r})\right)\cdot\partial\boldsymbol{r} = 0.$$

If we want to immunize $\boldsymbol{y}-\boldsymbol{x}$ against parallel shifts in the spot rate curve, then the second condition takes the form

$$D_{\boldsymbol{y}} = D_{\boldsymbol{x}}$$

whereas the identity

$$D_{\boldsymbol{y}}^{(1)} = D_{\boldsymbol{x}}^{(1)}$$

immunizes against steepening.

The portfolio must be rebalanced at the first payment and possibly earlier if the change of rate is considerable.

Assume that the portfolio $\boldsymbol{y}$ is composed by the bonds $\boldsymbol{b}_1, ..., \boldsymbol{b}_l$ and that $a_k$ is the number of the bond $\boldsymbol{b}_k$, $k = 1, ..., l$, in the portfolio;

$$\boldsymbol{y} = a_1\boldsymbol{b}_1 + \cdots + a_l\boldsymbol{b}_l.$$

Let $P_k$, $D_k$, etc., denote the prices, durations, etc., for $\boldsymbol{b}_k$. The equations above then take the form

$$a_1P_1 + \cdots + a_lP_l = P_{\boldsymbol{x}}$$

$$a_1P_1D_1 + \cdots + a_nP_lD_l = P_{\boldsymbol{x}}D_{\boldsymbol{x}}$$

$$a_1P_1D_1^{(1)} + \cdots + a_lP_lD_l^{(1)} = P_{\boldsymbol{x}}D_{\boldsymbol{x}}^{(1)}$$

or

$$v_1 + \cdots + v_l = 1$$

$$v_1 D_1 + \cdots + v_l D_l = D_{\boldsymbol{x}}$$

$$v_1 D_1^{(1)} + \cdots + v_l D_l^{(1)} = D_{\boldsymbol{x}}^{(1)},$$

where

$$v_k = \frac{a_k P_k}{P_{\boldsymbol{x}}}$$

is the weight of the bond $\boldsymbol{b}_k$ in the portfolio.

***Example 1.2***. Assume that the spot rates are as in Exercise 1.28 and that we want to imitate the cash stream $\boldsymbol{x} = (-P, 100, 100, 100, 100, 100)$ by means of the bonds in that exercise. Calculations show that $P = 430.47$, $D = 2.89$, and $D^{(1)} = 0.15$.

Let us first treat the case when we use only two of the three bonds and want to immunize against parallel shifts. If we want to have positive weights, then we cannot choose $A$ and $B$ since this gives a portfolio with duration $\geq 3.77 > D$ . We therefore must choose C and one of the others — $B$, for example.

We get the equations

$$v_B + v_C = 1$$
$$v_B D_B + v_C D_C = D$$

that have the solution

$$v_B = \frac{D - D_C}{D_B - D_C}, \qquad v_C = \frac{D_B - D}{D_B - D_C}.$$

The portfolio is therefore obtained by buying the bond B for $v_B P = 217.50$ EUR and C for $v_C P = 212.98$ EUR.

The number of the bonds B and C in the portfolio becomes $v_B P / P_B = 2.28$ and $v_C P / P_C = 2.35$, respectively. If we assume that the spot rate changes as in Fig. 1.2., then $\partial P_{\boldsymbol{x}} / P_{\boldsymbol{x}} = 81$ bp to be compared to $\partial P_{\boldsymbol{y}} / P_{\boldsymbol{y}} = 93$ bp.

EXERCISE 1.31. Make a portfolio by means of the bonds A and C which is immune relative to $\boldsymbol{x}$ against parallel shifts.

We have for this portfolio $\partial P_{\boldsymbol{y}} / P_{\boldsymbol{y}} = 91$ bp if the spot rate changes as in Fig. 1.2.

***Example 1.3***. Here we shall immunize also against steepening by using all three bonds: $A$, $B$, and $C$. In this case we have the equations

$$v_A + v_B + v_C = 1,$$

$$v_A D_A + v_B D_B + v_C D_C = D,$$

$$v_A D_A^{(1)} + v_B D_B^{(1)} + v_C D_C^{(1)} = D^{(1)},$$

which have the solution: $-0.27, 0.89, 0.38$. Here we thus need a short position in the bond A. We have for this portfolio $\partial P_y / P_y 94$ bp if the spot rate change as in Fig. 1.2. An impairment compared to the two portfolios in Example 1.2 and Exercise 1.31.

EXERCISE 1.32. Assume that there are zero-coupon bonds with optional time to maturity. Let $y$ consist of a number of *one* zero-coupon bond. Which condition must be satisfied to immunize against

(a) parallel shift.

(b) steepening.

If one immunizes the zero-coupon bond in (a) relative to $x$ against parallel shifts and the spot rate curve changes as in Fig. 1.2, then the relative change of price becomes 72 bp, to be compared to 81 bp for $x$.

EXERCISE 1.33. Assume that you shall pay 1 EUR in $T$ years and want secure the payment by buying a zero-coupon bond with time to maturity $T$ years. This does not exist, but there are zero-coupon bonds with times to maturity $T_1$ and $T_2$ years, where $T_1 < T < T_2$.

(a) Form a portfolio consisting of zero-coupon bonds with times to maturity $T_1$ and $T_2$ years. Determine the weights such that the portfolio has the same present value and duration as a zero-coupon bond with duration $T$.

(b) Let $\Delta_0(\boldsymbol{r})$ denote the difference between the value at time 0 of the portfolio and the zero-coupon bond with duration $T$ as a function of the rate $\boldsymbol{r}$.

Show that

$$\Delta_0(\boldsymbol{r} + \mathbf{1}\partial p) = \frac{d_T}{2}(T - T_1)(T_2 - T)(\partial p)^2 + O(|\partial p|^3).$$

The expression in (b) is thus positive for small $\partial p$. Assume that the interest rate at time $\partial t$ has changed to $\boldsymbol{r} + \mathbf{1}\partial p$ and that $\partial t$ is small compared to $\partial p$; thus the expression in (b) is an approximation of the difference of the values of the two portfolios at time $\partial t$.

## 1.3 NOTES

My main source here has been Luenberger (1998).

I have benefited from discussions with Harald Lang concerning the arbitrage theorem in Section 1.2.

The data in Section 2.4 are taken from Frye (1997).

# CHAPTER 2

# FURTHER FINANCIAL INSTRUMENTS

Here we shall introduce a few financial instruments that besides bonds can serve as building stones for portfolios. These are stocks, forwards, and options. Identities and inequalities for the prices of these are derived under the assumption that the assets are traded on an arbitrage-free market.

## 2.1 STOCKS

At any time during a trading day a stock usually has three prices: best bid, best ask, and last sale. As an example, we shall take the stock AstraZeneca on Stockholmsbörsen. At 13.30 July 12 2004 these prices were. Bid 325.00 SEK; Ask 325.50; Last sale 325.00. The difference between the ask and bid prices is called the *spread*. In this case it is thus 0.13% of the price.

These prices vary with time. The development of the last sale price during a part of the day July 12 2004 is shown in Table 2.1.

*Mathematical Asset Management.* By Thomas Höglund

**Table 2.1.** The development of tha price of AstraZeneca

| Time | Price | Number | Time | Price | Number |
|---|---|---|---|---|---|
| 13.19.47 | 326.50 | 50 | 13.29.49 | 325.00 | 300 |
| 13.19.56 | 326.50 | 1000 | 13.29.50 | 325.00 | 50 |
| 13.20.01 | 326.50 | 4000 | 13.30.42 | 325.50 | 150 |
| 13.20.19 | 326.50 | 850 | 13.31.33 | 325.50 | 200 |
| 13.21.37 | 327.00 | 2000 | 13.32.56 | 325.50 | 1600 |
| 13.24.45 | 327.00 | 1400 | 13.33.56 | 326.00 | 1000 |
| 13.25.45 | 327.00 | 5000 | 13.33.59 | 326.00 | 1000 |
| 13.25.48 | 327.00 | 2000 | 13.34.09 | 326.00 | 1000 |
| 13.27.45 | 326.00 | 13 | 13.34.12 | 326.00 | 950 |
| 13.27.56 | 326.00 | 850 | 13.34.17 | 326.00 | 1000 |
| 13.28.00 | 326.00 | 150 | 13.34.29 | 326.00 | 1000 |
| 13.28.05 | 326.00 | 1000 | 13.35.17 | 326.00 | 1000 |
| 13.28.07 | 326.50 | 200 | 13.35.56 | 326.00 | 200 |
| 13.28.43 | 326.00 | 12 | 13.37.25 | 325.50 | 1000 |
| 13.28.44 | 326.00 | 1000 | 13.37.27 | 325.00 | 2200 |
| 13.28.45 | 326.00 | 600 | 13.37.28 | 325.00 | 2000 |
| 13.28.46 | 326.00 | 1000 | 13.37.31 | 324.50 | 1450 |

It is typical for a liquid stock that the price moves up or down one unit with a few exceptional larger jumps. The unit in this case is 0.50 SEK, and larger jumps occur at 13.25.48–13.27.45 and between 13.28.46 and 13.29.49.

The price does not only vary with time but also with the volume. The order book of the above stock at 13.30 is shown in Table 2.2.

Therefore if we promptly want to sell 1950 stocks, we can do this for the price 325.00 per share; but if we sell

$$1950 + 5000 + 3745 + 2296 + 1000 = 13991$$

stocks, we will get

$$(1950 \times 325.00 + 5000 \times 324.50 + 3745 \times 324.00+$$

$$2296 \times 323.50 + 100 \times 323.00)/13991 = 324.16$$

per share. If we want to sell considerably more, this cannot be done at once without seriously affecting the price.

This phenomenon is more pronounced for less liquid stocks. The order book of a small company may be very thin. In extreme cases it is empty: There are neither any buyer nor any seller.

**Table 2.2.** The order book

| Bid Number | Price | Ask Number |
|---|---|---|
| | 327.50 | 13572 |
| | 327.00 | 5000 |
| | 326.50 | 6200 |
| | 326.00 | 8450 |
| | 325.50 | 2350 |
| 1950 | 325.00 | |
| 5000 | 324.50 | |
| 3745 | 324.00 | |
| 2296 | 323.50 | |
| 1000 | 323.00 | |

Above we have used the concept liquid stock without defining it. By a *liquid stock* we mean a stock that has small spread and for which the price is not considerably affected by the volume.

This somewhat vague definition thus depends on the volume. A stock that is liquid for one operator may be nonliquid for another.

### 2.1.1 Earnings, Interest Rate, and Stock Price

We shall here try to find a relationship between the price of a stock, $P$, the earnings per share, EPS, and the interest rate, $r$. The earnings per share is the yearly earnings of the company divided by the number of stocks. Below we shall write $E$ instead of EPS.

First consider an ideal world where nothing changes with time. The interest rate is always $r$ per year regardless of the duration, a certain company has every year the earnings $E$ per share, and all this is paid as dividend once a year. Let $P$ denote the price of the stock immediately after the payment of the dividend.

If one buys the stock and then sells it after the payment of the dividend the next year (for the same price since nothing has changed), then one gets the return $E/P$. For reasons of arbitrage, this will coincide with the interest rate. Therefore

$$P = \frac{E}{r}. \tag{2.1}$$

This identity thus holds in a stationary world without arbitrage opportunities.

The exercise below gives a crude idea of how small changes of earnings and interest rate affect the stock price.

EXERCISE 2.1. Assume that the above identity holds. Show that

$$\frac{\partial P}{P} = \frac{\partial E}{E} - \frac{\partial r}{r}.$$

In order to get a more realistic model, we shall assume that the price coincides with the present value of future earnings:

$$P = d_1 E_1 + d_2 E_2 + \cdots, \tag{2.2}$$

where $E_k$ is the earnings per share in $k$ years and $d_k$ is the $k-$year discount factor.

EXERCISE 2.2. Calculate $P$ when $E_k = E$ and $d_k = (\frac{1}{1+r})^k$ for $k = 1, 2, \ldots$.

The future earnings are not known, and different assumptions about these can give large differences in the price.

If the future earnings increase at least as fast as the rate, $d_k E_k \geq 1$, then the series diverges. Therefore, if we think this will be the case, no price is too high according to this model, provided that we have time to wait.

***Example 2.1.*** Assume we think we know that the earnings of a company will increase with $30\%$ per year the next 5 years: $E_k = (1.3)^k E$ for $k = 1, \ldots, 5$, where $E$ is the earnings the last year.

The present value of these earnings equals, with $d_k = (1/1.05)^k$, $10E$. If we in addition, discount unchanged earnings the following 5 years, $E_k = E_5$ for $k = 6, \ldots, 10$, then the present value will be $22E$. If we finally assume that these earnings will be unchanged for all time and discount this, we will get the present value $68E$. Here the multiples 10, 22 and 68 are rounded to the nearest integer.

If the expected growth $30\%$ fails to appear and the stock is revalued according to the assumption $E_k = E$ for all $k$, then the corresponding earnings multiples becomes 4, 8 and 20, respectively. The new price therefore is about $1/3$ of the old.

EXERCISE 2.3. Perform the above calculations.

The formulation of the result that corresponds to the result in Exercise 2.1 will be slightly simplified if $r_k$ stands for the continuous interest rate: $d_k = e^{-kr_k}$.

EXERCISE 2.4. Show that

$$\frac{\partial P}{P} = \sum_{k=1}^{\infty} p_k \frac{\partial E_k}{E_k} - \sum_{k=1}^{\infty} k r_k p_k \frac{\partial r_k}{r_k},$$

where $p_k = d_k E_k / P$ and therefore $p_1 + p_2 + \cdots = 1$.

In order to facilitate comparisons with the result of Exercise 2.1, we shall assume that

$$\frac{\partial E_k}{E_k} = \frac{\partial E}{E} \quad \text{and} \quad \frac{\partial r_k}{r_k} = \frac{\partial r}{r}$$

for all $k$. In this case we thus have

$$\frac{\partial P}{P} = \frac{\partial E}{E} - M\frac{\partial r}{r},$$

where

$$M = \sum_{k=1}^{\infty} k r_k p_k.$$

If $r_k = r$ and $E_k = E$ for all $k$, then

$$p_k = (1-d)d^{k-1}, \quad \text{and} \quad M = r(1-d)\sum_{k=1}^{\infty} k d^{k-1} = \frac{r}{1-d} = \frac{r}{1-e^{-r}} \approx 1.$$

On the other hand, if we discount considerable earnings in the distant future, then $M$ will be considerably larger. The prices of such stocks therefore are more sensitive to changes of the interest rate. In Example 2.1 we have $M = 2.7$.

We shall now see that formula (2.1) holds in general, provided that we define the earnings and the interest rate properly. Define $R$ by

$$\frac{1}{R} = d_1 + d_2 + \cdots.$$

This is thus the present value of one EUR per year forever.

EXERCISE 2.5. Show that if $d_k = (\frac{1}{1+r_k})^k$, then $\min r_k \leq R \leq \max r_k$. Hence in particular, $R = r$ if $r_k = r$ for all $k$.

Define $\bar{E} = Rd_1E_1 + Rd_2E_2 + \cdots$. This is an average value of future earnings since $Rd_1 + Rd_2 + \cdots = 1$. The valuation model (2.2) can now also be written

$$P = \frac{\bar{E}}{R}. \tag{2.3}$$

There are also other factors that influence the price. The bull market that culminated the spring 2000 cannot be explained by this formula. Except for psychological factors, a possible partial explanation is the increased amount of money on the market at that time.

**Summary 2.2.** *The price of a stock is determined to an essential degree by the relationship (2.3)*

*The prices of stocks that have discounted considerable earnings in the distant future are particularly sensitive to changes of the interest rate.*

## 2.2 FORWARDS

A *forward contract* is an agreement between two parties to purchase or sell an asset for a certain price at a certain future time, the *delivery date* or *maturity.* The buyer is said to have a *long position*, and the seller is said to have a *short position.* The *delivery price* is the price that applies at delivery. This price is chosen such that the initial value of the contract is zero. The only payment associated with the contract is at delivery. The *forward price* at a certain time is the delivery price of a contract signed at that time. In contrast to the delivery price of a specific contract, the forward price will vary with time.

Let $S_t$ denote the price of the asset (or commodity) at time $t$, and let $F$ stand for the delivery price of a contract written at $t = 0$ with delivery at $t = T$. We shall determine $F$, assuming that we are acting on a perfectly liquid market free of arbitrage. We shall also assume that there are no transaction costs and that the interest rate is the same regardless we borrow or lend.

First assume that the asset can be stored without costs. This is the case if, for example, the asset is a stock. In this case a short position can be hedged by purchasing the asset at time 0, and store it until $T$. This will give rise to the cash stream $(-S_0, F)$. On the other hand, a long position can be hedged by short selling of the asset at time zero, which gives the opposite stream $(S_0, -F)$. Therefore the present values of these must equal zero, $S_0 - d(0,T)F = 0$, and hence

$$F = S_0/d(0,T). \tag{2.4}$$

Here is a detailed argument: Assume that $F > S_0/d(0,T)$. At time 0: Borrow $d(0,T)F$ EUR, buy one asset, and take a short position in the forward. At time $T$ deliver the asset, receive $F$ EUR, and pay back the loan. You have obtained the stream $(d(0,T)F - S_0, 0)$ and hence a risk-free profit.

Assume that $F < S_0/d(0,T)$. At time 0: Sell one asset short, lend $d(0,T)F$, and take a long position in the forward. At time $T$ you will receive $F$ EUR from the borrower. Use these to buy back the asset. You have obtained the stream $(S_0 - d(0,T)F, 0)$ and hence a risk-free profit.

If we have storage costs, and the present value of these equals $C$, then a short position can be hedged by purchasing the commodity and paying the storage cost at time 0. This will give rise to the stream $(-S_0 - C, F)$.

In order to construct the opposite stream, one must not only be able to sell the commodity, but also to let storage space. If you think this is far-fetched, there is another factor to take into account, namely competition. For example, an alternative to take a long position is to borrow money, buy the commodity, and rent storage space.

Therefore, the party that holds the short position will not accept a lower delivery price than

$$F = \frac{S_0 + C}{d(0,T)}, \tag{2.5}$$

and competition will give a delivery price close to this.

If the storage costs are paid at $t_k$ and the cost at that time is $c_k$, $k = 0, 1, ..., n-1$, then

$$C = d(0, t_0)c_0 + \cdots + d(0, t_{n-1})c_{n-1}.$$

The forward contract entered into at $t = 0$ and with maturity $T$ has a value that varies with time $t$, $0 \leq t \leq T$, and the spot price, $S_t$. Let $f_t$ stand for the value at time $t$, and let $F_t$ stand for the forward price at $t$. Then

$$f_t = d(t, T)(F_t - F_0). \tag{2.6}$$

The reason for this is that if we enter into a forward contract at $t$, we will get the payoff $S_T - F_t$ at $T$ and hence the cash stream $(0, S_T - F_t)$. Alternatively, we can buy the old contract at $t$ and get the the cash stream $(-f_t, S_T - F_0)$. For reasons of arbitrage, these streams must have the same present values and hence (2.6) holds.

A *future* is very similar to a forward. The difference is irrelevant for our purposes, and therefore we shall not enter upon the subject.

## 2.3 OPTIONS

We shall here consider the two basic types of options: calls and puts. A call option gives the holder the right, but not the obligation, to buy the underlying asset for a certain price, the *strike price* or the *exercise price*, at a certain date, the *expiration date* or *maturity*. The other party that *writes* the option has the obligation, but not the right, to sell the asset. The holder of a put has the right to sell the asset for the strike price at maturity. Options that can be exercised at any time up to the expiration date are called *American*, whereas options that can be exercised only at the expiration date are called *European*. The holder of an option is said to have a *long position*, whereas the writer has a *short position*.

The underlying asset can be almost anything, but we shall assume that the options are traded on a liquid market free of arbitrage. Therefore only assets such as stocks, stock indices, foreign currencies, and a few others are possible. Here we shall assume that the underlying asset is a stock, and unless otherwise is stated, this stock pays no dividends.

### 2.3.1 European Options

We shall write $K$ for the strike price and $T$ for the time to maturity. The value at time $t, 0 \leq t \leq T$, of a call and a put will be denoted by $C_t$, and $P_t$, respectively. Thus

$$C_T = \max(0, S_T - K), \text{ and } P_T = \max(0, K - S_T).$$

Here $S_t$ denotes the price of the stock at time $t$. Sometimes we will also write $C_t(K)$ or $C_t(K, T)$ instead of $C_t$ when we think this is preferable.

How to calculate the exact value of of these options will be described in Chapter 5. We shall here derive a few general properties assuming that the options are traded on a perfectly liquid market without arbitrage opportunities. We shall also assume that there are no transaction costs and that the (continuous) interest rate is constant and equals $r$ regardless we borrow or lend.

**Proposition 2.3.**

(a) *A call is a decreasing function of the strike price.*

(b) $e^{-rT}K + C_0(K)$ *is an increasing function of the strike price* $K$.

(c) $C_0(K) \geq \max(0, S_0 - e^{-rT}K)$.

(d) *A call is a convex function of the strike price.*

Any convex function is continuous, and hence it follows from (d) that a call is a continuous function of the strike price.

The proof of (a) and (b) will be left as exercises.

*Proof.* (c) If $C_0 < 0$, then the acquisition of a call will give a certain strictly positive income at time zero, along with an income at maturity which is at least 0, contradicting the assumption of no arbitrage. Hence $C_0(K) \geq 0$.

Assume that

$$C_0 < S_0 - e^{-rT}K.$$

Sell a stock (which you do not have to own). Buy a call and keep the income

$$I = S_0 - C_0 > e^{-rT}K$$

in the cash account. Wait until maturity. Then buy back the stock (for $S_T$ EUR). The cash has now increased to $Ie^{rT}$, and the portfolio value equals

$$F_T = Ie^{rT} + C_T - S_T = F_0 e^{rT} + \max(0, S_T - K) - S_T >$$

$$K + \max(0, S_T - K) - S_T = \max(K - S_T, 0) \geq 0$$

at time $T$.

The portfolio value at time 0 equals 0, and hence the assumption gives rise to arbitrage, a contradiction.

(d) By (a), $C$ is decreasing, and hence it suffices to show that

$$C(\frac{K_1 + K_2}{2}) \leq \frac{C(K_1) + C(K_2)}{2}.$$

Assume that $K_1 < K_2$. Write at time 0 two calls with strike price $(K_1 + K_2)/2$, and buy a call with strike price $K_1$ and another with strike price $K_2$. This will give the income

$$I = 2C_0(\frac{K_1 + K_2}{2}) - C_0(K_1) - C_0(K_2)$$

at time 0. The value of this portfolio at maturity, $F_T$, satisfies

$$F_T - Ie^{rT} = C_T(K_1) + C_T(K_2) - 2C_T(\frac{K_1 + K_2}{2}) =$$

$$\max(0, S_T - K_1) + \max(0, S_T - K_2) - 2\max(0, S_T - \frac{K_1 + K_2}{2}) =$$

$$\begin{cases} 0 & \text{if } S_T \leq K_1 \text{ or } S_T > K_2 \\ S_T - K_1 & \text{if } K_1 < S_T \leq \frac{K_1+K_2}{2} \\ K_2 - S_T & \text{if } \frac{K_1+K_2}{2} < S_T \leq K_2 \end{cases},$$

and hence $F_T \geq Ie^{rT}$. Therefore $I \leq 0$. □

EXERCISE 2.6. Show (a) and (b) of the above proposition.

We have above compared calls with the same maturity. The next proposition shows the effect of different expiration dates and strike prices.

**Proposition 2.4.** *If $T_1 \leq T_2$ and $e^{-rT_1} K_1 \geq e^{-rT_2} K_2$, then*

$$C_0(K_1, T_1) \leq C_0(K_2, T_2).$$

Note that the assumption and hence the conclusion holds if $T_1 \leq T_2$ and $K_1 = K_2$, or if $T_1 = T_2$ and $K_1 \geq K_2$,

*Proof.* Assume that $C_0(K_2, T_2) < C_0(K_1, T_1)$. Buy the call with strike price $K_2$ and write the call with strike price $K_1$, and keep the income $I = C_0(K_1, T_1) - C_0(K_2, T_2) > 0$ in the cash account. The value of the portfolio at $T_1$ then satisfies

$$F_{T_1} = Ie^{rT_1} + C_{T_1}(K_2, T_2) - C_{T_1}(K_1, T_1) > C_{T_1}(K_2, T_2) - \max(0, S_{T_1} - K_1) \geq$$

$$\max(0, S_{T_1} - e^{-r(T_2 - T_1)}K_2) - \max(0, S_{T_1} - K_1).$$

Here we used Proposition 3a.

It follows that

$$e^{-rT_1}F_{T_1} > \max(0, e^{-rT_1}S_{T_1} - e^{-rT_2}K_2) - \max(0, e^{-rT_1}S_{T_1} - e^{-rT_1}K_1) \geq 0$$

since

$$e^{-rT_1}K_1 \geq e^{-rT_2}K_2.$$

We can therefore at time $T_1$ sell the option with maturity $T_2$, and pay $\max(0, S_{T_1} - K_1)$. It remains $Ie^{rT_1}$ in cash, and we have no obligations at $T_2$. □

The following important identity relates the prices of a put and a call.

**Put-Call Parity**

$$e^{-rT}K + C_0(K) = S_0 + P_0(K). \tag{2.7}$$

*Proof.* We can acquire a portfolio which at time zero gives us an income that equals the difference of the two sides of (2.7),

$$I = e^{-rT}K + C_0(K) - S_0 - P_0(K).$$

Buy a stock and a put, and write a call and borrow the present value of the strike price. The value of this portfolio at maturity equals

$$F_T = Ie^{rT} - K - \max(0, S_T - K) + S_T + \max(0, K - S_T) = Ie^{rT}.$$

Hence $I \leq 0$.

We can also at time zero acquire the portfolio which give the income $-I$ at time 0. (Sell a stock and write a put, and buy a call and lend the present value of the strike price.) Therefore $-I \leq 0$ and hence $I = 0$.

EXERCISE 2.7. Use options to construct a forward contract on a stock.

EXERCISE 2.8. Show that the put-call parity takes the form

$$e^{-rT}K + De^{-rt} + C_0(K) = S_0 + P_0(K)$$

in the case when the stock pays the dividend $D$ at time $t$, $0 < t < T$.

Applying the put-call parity to (b), (a), (c), and (d) of Proposition 3, we get the next proposition.

**Proposition 2.5.**

(a) *A put is an increasing function of the strike price.*

(b) $P_0(K) - e^{-rT}K$ *is a decreasing function of the strike price* $K$.

(c) $P_0(K) \geq \max(0, Ke^{-rT} - S_0)$

(d) *A put is a convex function of the strike price.*

EXERCISE 2.9. Perform the details of the above proof.

### 2.3.2 American Options

Compared to a European option, an American has the advantage that it is possible to exercise it at any time $t \leq T$. Therefore with obvious notation we have

$$C^{\text{Am}}(K) \geq C^{\text{Eu}}(K) \text{ and } P^{\text{Am}}(K) \geq P^{\text{Eu}}(K).$$

We shall show that equality holds in the first of these inequalities but not in the second.

**Proposition 2.6.** *It does not pay to exercise an American call before maturity, but it is better to sell it.*

*Proof.* It follows from Proposition 3 that

$$C_0^{\text{Am}}(K) \geq \max(0, S_0 - e^{-rT}K) > S_0 - K$$

for all $S_0$, and the latter is what we get if we exercise the call. □

In contrast to this result it may pay to exercise a put.

**Proposition 2.7.** *If*

$$S_0 < K(1 - e^{-rT}), \tag{2.8}$$

*then it is better to exercise the put and keep the money in the cash account compared to waiting until maturity.*

*Proof.* If we exercise the put, then the value of the portfolio at maturity equals

$$(K - S_0)e^{rT} > K.$$

Otherwise it equals

$$\max(0, K - S_T) \leq K.$$

□

If the converse to the inequality (2.8) holds, then it had been better to wait if it turns out that

$$e^{-rT}S_T < S_0 - K(1 - e^{-rT}),$$

but not otherwise.

### 2.3.3 Option Strategies

By combining long and short positions of calls, puts, stocks, and cash, it is possible to construct a wide variety of portfolios. Consider options with the same maturity, $T$. If all strike prices were available, then it would be possible to construct a portfolio whose payoff (value at maturity) is $f(S_T)$. Here $f$ is any continuous function whose graph consists of a finite number of pieces of straight lines. (Here we have neglected the initial investment.)

Thus, for example, if we buy a call with strike price $K_1$ and write a call with strike price $K_2 > K_1$, the payoff will be

$$\begin{array}{cl} 0 & \text{if } S_T \leq K_1, \\ S_T - K_1 & \text{if } K_1 < S_T \leq K_2, \\ K_2 - K_1 & \text{if } S_T > K_2. \end{array}$$

The profit will therefore be

$$\begin{array}{cl} C_0(K_2) - C_0(K_1) & \text{if } S_T \leq K_1, \\ C_0(K_2) - C_0(K_1) + S_T - K_1 & \text{if } K_1 < S_T \leq K_2, \\ C_0(K_2) - C_0(K_1) + K_2 - K_1 & \text{if } S_T > K_2. \end{array}$$

With a proper choice of the strike prices, this portfolio (which is called a *bull spread*) will give a positive payoff if the stock price rises, and the loss will be at most $C_0(K_1) - C_0(K_2)$. If we instead believe in a decrease in the stock price, we can construct a *bear spread*: Write a call with strike price $K_1$ and buy one with strike price $K_2$. In the case where the options are European, we can use the put-call parity and construct the portfolios with puts instead of calls. To get a bull spread, we thus buy a put with strike price $K_1$ and write another with strike price $K_2$ (and keep $K_2 - K_1$ in the cash account).

EXERCISE 2.10. Verify the details of the reasoning above.

A few other examples are given in the next exercise.

EXERCISE 2.11. Construct portfolios with the following payoffs.

(a) (Butterfly spread) 0 if $S_T \leq K_1$, $S_T - K_1$ if $K_1 < S_T \leq K_2$, $K_3 - S_T$ if $K_2 \leq S_T < K_3$, 0 if $K_3 < S_T$. Here $K_1 < K_3$ and $K_2 = (K_1 + K_3)/2$.

(b) (Straddle) $K - S_T$ if $S_T \leq K$, $S_T - K$ if $S_T > K$.

(c) (Strangle) $K_1 - S_T$ if $S_T \leq K_1$, 0 if $K_1 < S_T \leq K_2$, $S_T - K_2$ if $K_2 < S_T$. Here $K_1 < K_2$.

In the remainder of this section we shall consider three strategies in some detail. We shall assume that we have a stock and, based on this construct, three different option portfolios.

***Insurance with Puts. (Protective Put)*** By buying a put with strike price $K$ and time to expiration $T$, one will be sure to be able to sell the stock for at least $K$ EUR. The development of the portfolio depends on the way the purchase of the option is financed.

If this is financed with cash, the portfolio value at $t$ becomes

$$F_t = S_t + P_t(K) - P_0(K)e^{rt}.$$

In this case we thus have

$$F_T = \max(K, S_T) - P_0(K)e^{rT}.$$

Normally, stock options are American. If this is the case, then it is possible to exercise the option (i.e., sell the stock for $K$ EUR) at any time $t \leq T$ and hence $F_t \geq K - P_0(K)e^{rt}$ for all $t \leq T$.

Instead of financing the purchase of the option with cash, one can sell $x$ stocks and hedge the remaining stocks with $1-x$ puts. Here consequently $xS_0 = (1-x)P_0(K)$ and the value of the portfolio at $t$ becomes $F_t = (1-x)(S_t + P_t(K))$, that is,

$$F_t = \frac{S_t + P_t(K)}{1+p}, \quad \text{where } p = P_0(K)/S_0. \tag{2.9}$$

In this case

$$F_T = \frac{\max(K, S_T)}{1+p}.$$

If the option is American, then $F_t \geq K/(1+p)$ for all $t \leq T$.

EXERCISE 2.12. As alternatives to this portfolio, one can choose to keep all the money, $S_0$, in the cash account or to buy the stock for the entire amount.

Assume that $e^{-rT}K < S_0$. For which values of $S_T$ is this portfolio worth at least as much at time $T$ as if you had put all the money (a) in cash? (b) in the stock? (c) Which is the best of these alternatives?

EXERCISE 2.13.

(a) Assume that the options are European. Construct a portfolio consisting only of calls and puts and cash (but no stock) and which has the same value as $S_t + P_t(K)$ for all $0 \leq t \leq T$.

(b) Assume that the options are American. Discuss advantages and disadvantages from the point of view of risk with the two portfolios.

***Finance the Insurance with a Short Position in a Call*** If

$$e^{-rT}K < S_0,$$

the above portfolio insurance can also be financed by writing a call with an appropriately chosen strike price $H$.

EXERCISE 2.14. Assume that the options are European and that

$$e^{-rT}K < S_0.$$

(a) Show that $P_0(K) < C_0(K)$.

(b) Since $C_0(x) \to 0$ as $x \to \infty$, there is a $H > K$ such that $C_0(H) = P_0(K)$. Show that

$$e^{-rT}H > S_0.$$

We shall assume that such an $H$ exists even if the put is American. The value of the portfolio at $t \leq T$ consequently equals

$$F_t = S_t + P_t(K) - C_t(H). \tag{2.10}$$

In this case

$$F_T = \begin{cases} K & \text{if } S_T \leq K, \\ S_T & \text{if } K < S_T \leq H, \\ H & \text{if } H < S_T, \end{cases}$$

that is, $F_T = K + (S_T - K)_+ - (S_T - H)_+$.

EXERCISE 2.15. Verify this.

EXERCISE 2.16. The same as in Exercise 2.12 but with this portfolio.

EXERCISE 2.17. The same as in Exercise 2.13 but with this portfolio.

***A Long Position in a Stock and a Short Position in a Call (Covered Call).*** This strategy is systematically used by some portfolio managers. One writes a call with a strike price, $H$, above the present stock price. The advantage is that one receives $C_0(H)$ EUR. The "risk" is that the stock price increases above $H$ and the option is exercised.

If the income from the call is kept in cash the value of the portfolio at $t \leq T$ equals

$$F_t = S_t - C_t(H) + C_0(H)e^{rt},$$

and thus

$$F_T = \min(H, S_T) + C_0(H)e^{rT}.$$

By writing $1 + x$ calls and buying $x$ stocks, where $xS_0 = (1 + x)C_0(H)$, one can reinvest the income in the stock. The value of the portfolio then becomes $F_t = (1 + x)(S_t - C_t(H))$, that is,

$$F_t = \frac{S_t - C_t(H)}{1 - c}, \quad \text{where } c = C_0(H)/S_0. \tag{2.11}$$

In this case

$$F_T = \frac{\min(H, S_T)}{1 - c}.$$

EXERCISE 2.18. The same as in Exercise 2.12 but with the portfolio

$$(S_t - C_t(H))/(1 - c)$$

in the case when $e^{-rT}H > S_0$.

EXERCISE 2.19. The same as in Exercise 2.13 but with the portfolio $S_t - C_t(H)$.

***Comparison of the Strategies*** Choose $K$ such that $e^{-rT}K < S_0$ and let $H$ be determined by $C_0(H) = P_0(K)$. We shall here compare the three strategies

1. $(S_t + P_t(K))/(1 + p)$
2. $S_t + P_t(K) - C_t(H)$
3. $(S_t - C_t(H))/(1 - c)$.

EXERCISE 2.20.

(a) Plot the values of these portfolios at maturity $T$ as functions of $S_T$.

(b) For each value of $S_T$, which portfolio value is the largest?

(c) For each value of $S_T$, which portfolio value is the smallest?

## 2.4 FURTHER EXERCISES

EXERCISE 2.21.

(a) The price of copper is today 1739 USD per ton. Determine the delivery price of a forward contract for delivery of one ton copper in three months if the storage cost is 2.10 USD per month and ton, and the yearly return rate is $4\%$. The storage cost is paid in the beginning of each month.

(b) What is the value of this contract in one month if the spot price then equals 1760 USD?

The next exercise is an example where the "storage cost" is negative.

EXERCISE 2.22. A stock pays dividends once a year. The price of the stock is today, immediately after the payment of dividends, 10 EUR. It is known that the dividends for the following three years will be 0.5 EUR per year. The interest rate is flat and the yearly return rate is $4\%$.

Determine the delivery price of a forward for delivery of a stock just after the payment of dividends in three years.

EXERCISE 2.23. You intend to acquire a stock in two years. Today the price of the stock is 10 EUR. The continuous interest rate is $4.77\%$ per year, regardless of its duration. The stock pays no dividends.

In order to be able to control the cost, you consider two alternatives: to buy a call or take a long position in a forward. The price of a call with strike price 11 EUR and time to maturity 2 years equals 2.51 EUR.

(a) Determine the delivery price of a forward for delivery of one stock in two years.

(b) Each of the two procedures gives rise to a cash stream. Determine these. Decide for which values of $S_2$ the cash stream determined by the call has a higher present value than the other. Here $S_2$ stands for the price of the stock in two years.

## 2.5 NOTES

More material on this subject can be found in Hull (1989).

# CHAPTER 3

# TRADING STRATEGIES

In this chapter we show how to manage risk by trading. We shall sell or buy a quantity of an asset, depending on what happens to the price. No assumption whatsoever about the stochastic behavior of the asset prices will be made. We will, however, require that asset prices are reasonably continuous.

If one assumes that stock prices really are continuous and that stocks can be traded in arbitrarily small amounts, one gets a neat asymptotic approximation by trading infinitely often. This result will be used later in Chapter 5 to derive the Black–Scholes formula.

## 3.1 TRADING STRATEGIES

Let $S_t$ denote the price of a unit of a stock (or asset) at time $t$. We shall consider portfolios of the form

$$P_t = c_t + a_t S_t.$$

Here $P_t$ is the portfolio value, $c_t$ the cash holding, and $a_t$ the number of assets at time $t$. We shall assume that the interest rate equals 0, or rather we consider the present values of the above quantities.

Let us start with a simple example. Assume that $S_0 = 20$ EUR. Let $t_1 > 0$ denote the first moment at which the stock costs either 22 or 18 EUR, and let $t_2 > t_1$ denote the first moment when the price of the stock has increased or decreased another 2 EUR. This gives the following *binary tree*:

$$\begin{array}{cccc} t: & 0 & t_1 & t_2 \\ \\ & & & 24 \\ & & 22 & \\ S_t: & 20 & & 20 \\ & & 18 & \\ & & & 16 \end{array}$$

Here we shall consider a number of portfolios that at $t = 0$ are worth 20 EUR.

**Portfolio 0.** Keep the money in the cash account; $P_t = 20$ for $t = 0, t_1, t_2$.

**Portfolio 1.** Buy a unit of the stock for 20 EUR at $t = 0$. Do no more; $P_t = S_t$ for $t = 0, t_1, t_2$. This portfolio will thus have the same development as the stock.

**Portfolio 2.** Buy half a unit of the stock for 10 EUR at $t = 0$. Do no more; $P_t = 10 + \frac{1}{2}S_t$ for $t = 0, t_1, t_2$. This portfolio will have the following development:

$$\begin{array}{cccc} & & & 22 \\ & & 21 & \\ P_t: & 20 & & 20 \\ & & 19 & \\ & & & 18 \end{array}$$

Common to the above portfolios is that we do not change them during the journey (i.e., at $t = t_1$). By changing the portfolio at $t = t_1$, we can steer for a target given in advance. (The initial value will, however, depend on the target.)

**Portfolio 3.**

$$\begin{array}{cccc} & & & 23 \\ & & 21 & \\ P_t: & 20 & & 19 \\ & & 19 & \\ & & & 19 \end{array}$$

The portfolio is given by the following trading strategy:

$t = 0$. Buy $\frac{1}{2}$ of a unit of the stock; $P_0 = 10 + \frac{1}{2}S_0$.

$t = t_1$. If $S_{t_1} = 22$, buy another $\frac{1}{2}$ unit of the stock; $P_{t_1} = -1 + S_{t_1}$. (We thus have to borrow 1 EUR.) If $S_{t_1} = 18$, sell all stocks $P_{t_1} = 19$.

The number of stock units is given by the following binary tree:

$$
a_t : \begin{array}{ccc} & & \cdot \\ & 1 & \\ \frac{1}{2} & & \cdot \\ & 0 & \\ & & \cdot \end{array}
$$

And the cash holding by

$$
c_t : \begin{array}{ccc} & & \cdot \\ & -1 & \\ 10 & & \cdot \\ & 19 & \\ & & \cdot \end{array}
$$

How to construct this portfolio? Start from the right, that is, at $t = t_2$. At the top, for example, we have

$$
\begin{array}{cc} t_1 & t_2 \\ & \\ & 23 \\ ? & \\ & 19 \end{array}
$$

We must be in a position at $t_1$ such that we are able to reach the target at $t_2$. Assume that $P_{t_1} = c + aS_{t_1}$. Then we must have $c + a24 = 23$ and $c + a20 = 19$, that is, $a = 1$ and $c = -1$. The portfolio value at $t_1$ equals $-1 + 1 \times 22 = 21$. In the same way the other portfolio value at $t_1$ is calculated. Once this is done, we can proceed backward to $t = 0$.

The construction of a trading strategy is thus reduced to the following: Assume that we have the stock development:

$$
\begin{array}{cc} & s_u \\ s & \\ & s_d \end{array}
$$

where $s_d < s < s_u$, and that the portfolio looks like

$$
\begin{array}{cc} & f_u \\ f & \\ & f_d \end{array}
$$

Here $f_u$ and $f_d$ are given in advance, and $f = c + as$, where $c$ and $a$ are to be determined by the system: $c + as_u = f_u, c + as_d = f_d$.

EXERCISE 3.1. Show that the system has the solution

$$a = \frac{f_u - f_d}{s_u - s_d}, \quad c = \frac{s_u f_d - s_d f_u}{s_u - s_d}.$$

EXERCISE 3.2. Show that

$$f = qf_u + (1 - q)f_d, \quad \text{where } q = \frac{s - s_d}{s_u - s_d}.$$

Observe that it is not necessary to solve the equation, but it is often simpler to use the above formulas to calculate $q$ (which in the example equals $1/2$) and then the portfolio values (in the example, $f$ equals the mean value of $f_u$ and $f_d$). After that the numbers of the stock, $a$, are calculated, and last the cash holding. A simple way to calculate the cash holding is to use the formula $c = f - as$.

EXERCISE 3.3. Consider the same stock development as in the portfolios 0–3. Calculate the development of the portfolio value, the number of stock units, and the cash for the trading strategy that has the portfolio value $f(S_{t_2})$ at time $t_2$. Here $f(16) = 23, f(20) = f(24) = 19$.

In this example, negative stock holdings occur (short selling or shorting), which means that one borrows the stock and sells it (in order to be able to buy it back later for a price which hopefully is lower).

EXERCISE 3.4. As in Exercise 3.3 but with $f(16) = f(20) = 0$ and $f(24) = 80$.

In this case both the pledge and the risk is high.

The following example illustrates how the above trading strategies might work in real life.

***Example 3.1.*** Framfab

Framtidsfabriken, later renamed Framfab, was launched on Stockholmsbörsen on June 23, 1999. The stock proved to be fierce: It closed at the price 19.50 SEK on the first day. It reached a peak price after 180 trading days and closed at $323 = 16.56 \times 19.50$, SEK. On November 23, 2000 the price was for the first time below 19.50, and the stock closed at $19.00 = 0.97 \times 19.50$.

We shall use the trading strategy of Portfolio 3 for this stock during the above mentioned period. We shall assume that the stock is traded only at the daily closing prices. Therefore we cannot expect to trade exactly at the levels of the binary tree, but we will trade at the first time points the levels are reached or passed.

At time $t_2$ we begin again by putting half of the portfolio value in the cash account and half in the stock. We have disregarded transactions costs like spread and brokerage, but also the interest rate.

The development of the stock and Portfolio 3 are shown in Fig. 3.1. The stock development is unbroken and the portfolio is dashed. The scale is logarithmic. Each unit on the vertical axis corresponds to a doubling/halving.

The value of the stock on the final day was $0.97$ of the initial value whereas the portfolio value was $2.20$, $+120\%$. The volatility of the stock (i.e., the standard deviation of the daily returns) was $120\%$ on a yearly basis, whereas the volatility of the portfolio was $69\%$.

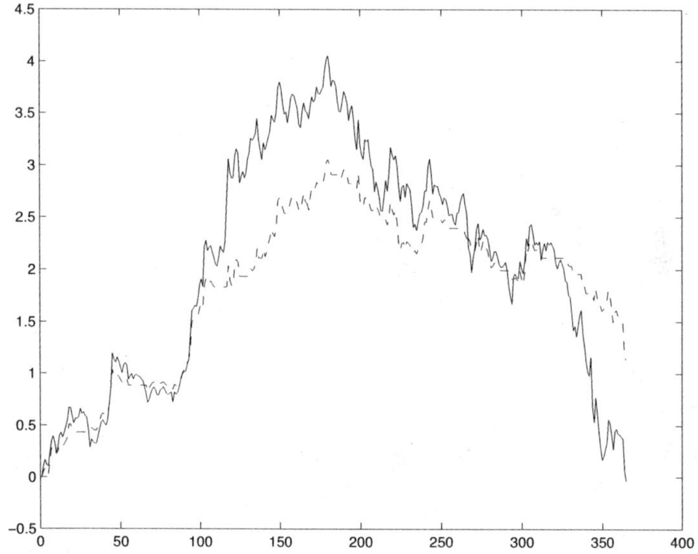


**Figure 3.1.** The development of Framfab (unbroken) and Portfolio 3 (dashed) during the period June 1999–November 2000. Logarithmic scale.

Assume that on November 23, 2000 we gave up believing in a growth of the stock and switched to the short-selling strategy of Exercise 3.3. The developments of this portfolio and the stock during the period November 23, 2000 – June 20, 2001 are shown in Fig. 3.2. The scale is logarithmic. The stock decreased during the period from 19 to $1.01 = 0.053 \times 19$. The portfolio value increased with a factor $1.42$, $+42\%$. Both the stock and the portfolio had high volatilities — $1.97$ and $1.88$, respectively — on a yearly basis. A reason for the high portfolio volatility is that we only trade at the closing prices and that the portfolio certain days moves considerably more than $5\%$ before it will be rebalanced. This holds in particular between the days 110, 111, and 112 and the days 122 and 123.

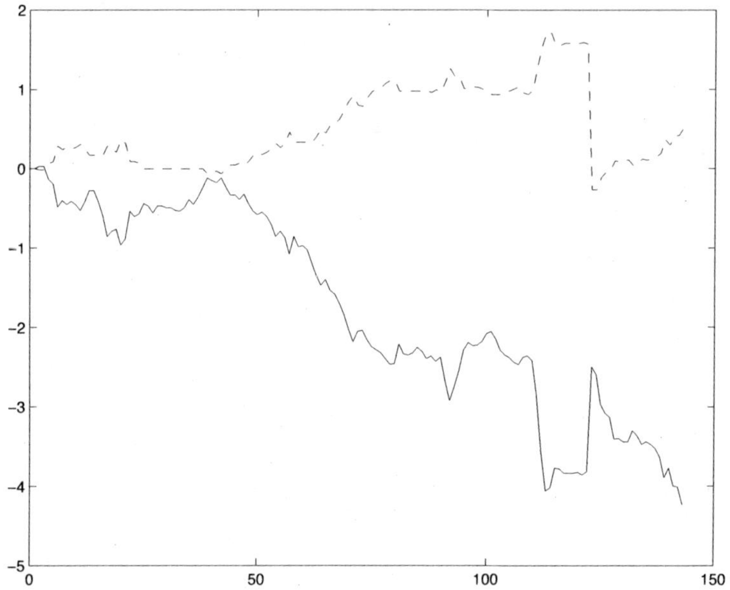


**Figure 3.2.** The development of Framfab and the portfolio of Exercise 3.3 during the period June 1999–November 2000. Logarithmic scale.

If we had used the trading strategy of Exercise 3.4 the portfolio would have been worthless fairly soon regardless of where we had started. □

One can, of course, consider more than three time points. The stock is traded at the levels given by a binary tree of the form

$$
\begin{array}{cccccc}
t: & 0 & t_1 & t_2 & \cdot\ \cdot & t_n \\
 & & & & & b_{n,n} \\
 & & & & \cdot & \\
 & & & \cdot & & \\
 & & & b_{2,2} & & \\
 & & b_{1,1} & & & \\
S_t: & b_{0,0} & & b_{2,1} & & \\
 & & b_{1,0} & & & \\
 & & & b_{2,0} & & \\
 & & & \cdot & & \\
 & & & & \cdot & \\
 & & & & & b_{n,0}.
\end{array}
$$

Here $b_{0,0} = S_0$ and $b_{k+1,j} < b_{k,j} < b_{k+1,j+1}$ for $j = 0, 1, ..., k$, $k = 0, 1, ..., n-1$.

A function $f$ is called the *target function* if the portfolio value equals $f(S_{t_n})$ at $t_n$.

In the next exercise you will compare trading strategies that have the same target functions as a call and a put, respectively.

EXERCISE 3.5. A stock is traded at the levels of the binary tree $b_{k,j}$ $j = 0, ..., k$, $k = 0, 1, 2, 3, 4$, where $b_{0,0} = S_0 = 10$ and $b_{k+1,j+1} = b_{k,j} + 1$, $b_{k+1,j} = b_{k,j} - 1$, $j = 0, ..., k$, $k = 0, ..., 3$. Calculate the development of the portfolio value, the number of stock units, and the cash for the trading strategy that has the target function $f$ in the following two cases.

(a) European call with the strike price 12: $f(s) = \max(0, s - 12)$. What is the price of the option, that is, the value of the portfolio at $t = 0$?

(b) European put with the strike price 8: $f(s) = \max(0, 8 - s)$.

(c) Assume that you hold a unit of the stock, buy the put in (b), and write the call in (a). Determine the development of the portfolio, the number of stock units, and the cash.

Exercise 3.5c thus gives an idea of the trading strategy that lies behind the option strategy "Finance the insurance with a short position in a call" in Section 2.3.3.

EXERCISE 3.6. A stock is traded at the levels of the binary tree $b_{k,j}$ $j = 0, ..., k$, $k = 0, 1, 2, 3, n$, where $b_{0,0} = S_0$ and $b_{k+1,j+1} = b_{k,j} + h$, $b_{k+1,j} = b_{k,j} - h$, $j = 0, ..., k$, $k = 0, ..., n$. Let $F_k(S_{t_k})$ stand for the portfolio value at $t = t_k$ for the trading strategy that has the target function $f$. Here $S_0 > nh$ and $f$ is a given function. Show that

$$F_{n-k}(s) = 2^{-k} \sum_{j=0}^{k} \binom{k}{j} f(s + (k - 2j)h).$$

### 3.1.1 Model Assumptions

When we constructed the above trading strategies, we made the implicit assumption that the traded asset satisfies two conditions (divisibility and liquidity) that are satisfied even for certain assets that are not stocks, but that are not satisfied for certain stocks.

It must be possible to trade the asset in parts. Exactly how small is determined by the trading strategy.

**Divisibility assumption.** *It is possible to trade the asset in sufficiently small parts.*

We assumed in the introductory example that if the value of the stock decreases from 20 EUR at $t = 0$ to 16 EUR at $t = t_2$, then it is possible to to trade it in between at the level 18 EUR.

**Liquidity assumption.** *It is possible to trade the asset at all levels of the binary tree.*

The price of a liquid stock that costs 20 EUR, say, moves as a rule up or down in steps of 5 cents and the spread is often of the same size. The above assumption seems reasonable if the distance between the levels is some multiples of the stock price multiplied by $\frac{0.05}{20} = 0.25\%$. But there is more than one instance of liquid stocks that have fallen $30\%$ in a few minutes. So one has to be alert. The reason for the considerable fluctuations of the portfolio value in Fig. 3.2 is that the liquidity condition was not always satisfied, but the stock price landed far outside the binary tree. Hence this holds at least if we only trade at the end of the day.

### 3.1.2 Interest Rate

If the cash account yields the continuous interest $r > 0$, then

$$\bar{P}_t = \bar{c}_t + a_t \bar{S}_t,$$

where $\bar{P}_t = e^{-rt} P_t$ denotes the present value of $P_t$, and correspondingly for $\bar{c}_t$ and $\bar{S}_t$. In the same way as above, we can thus steer the present value of the portfolio by rebalancing the portfolio at the moments when the present value of the stock price reaches the levels of the binary tree.

### 3.1.3 Exotic Options

All the portfolios we have treated have one thing in common: The final value $f(S_{t_n})$ depends only on the value of the stock at the final moment $t_n$. There are, however, financial derivatives called exotic options whose values depend on the value of the stock at several time points. We shall give an example to illustrate how to modify the above method in this case.

Consider the same stock development as in the introductory example. We shall construct a portfolio whose value at $t_2$ equals $\max(S_0, S_{t_1}, S_{t_2})$. If $S_{t_2} = 20$, then we must know which of the two values 22 and 18 the stock had at $t_1$. Therefore we write the binary tree in the following way:

$$
\begin{array}{cccc}
 & & & 24 \\
 & & 22 & \\
 & & & 20 \\
S_t : & 20 & & \\
 & & & 20 \\
 & & 18 & \\
 & & & 16
\end{array}
$$

The portfolio value will have the development

$$
\begin{array}{lccc}
 & & & 24 \\
 & & 23 & \\
 & & & 22 \\
P_t: & 21.5 & & \\
 & & & 20 \\
 & & 20 & \\
 & & & 20
\end{array}
$$

The rightmost column thus equals the maximal value of the stock. We have obtained the remaining values by working backward in the tree in the same way as before.

EXERCISE 3.7. Norm this portfolio in such a way that the initial value becomes 20. Also calculate the developments of the number of stock units and the cash.

EXERCISE 3.8. Consider the same stock development as in the above example. Determine the trading strategy that gives the portfolio value $(S_0 + S_{t_1} + S_{t_2})/3$. Compare the development of this portfolio with that of Exercise 3.7.

## 3.2 AN ASYMPTOTIC RESULT

### 3.2.1 The Model of Cox, Ross, and Rubinstein

So far we have only rebalanced the portfolio when the differences, $S_{t_k} - S_{t_{k-1}}$, reach given levels. A reason for this is that the calculations becomes very simple.

In the sequel we shall instead rebalance the portfolio when the quotients $S_{t_k}/S_{t_{k-1}}$ reach given levels. Thus consider a binary tree of the form

$$
\begin{array}{lcccccc}
t: & 0 & t_1 & t_2 & \cdot & \cdot & t_n \\
 & & & & & & S_0u^n \\
 & & & & & \cdot & \\
 & & & & \cdot & & \\
 & & & S_0u^2 & & & \\
 & & S_0u & & & & \\
S_t: & S_0 & & S_0ud & & & \\
 & & S_0d & & & & \\
 & & & S_0d^2 & & & \\
 & & & & \cdot & & \\
 & & & & & \cdot & \\
 & & & & & & S_0d^n,
\end{array}
$$

where $d$ and $u$ are positive numbers such that $d < 1 < u$. The interest rate equals 0 as before.

EXERCISE 3.9.

(a) Calculate $q$ expressed in terms of $d$ and $u$.

(b) Calculate $q$ expressed in $u$ in the case when $d = 1/u$.

(c) What is $q$ if $d + u = 2$? (That is if $d = 1 - b,\ \ u = 1 + b$ for some $0 < b < 1$.)

EXERCISE 3.10. Consider the case when $n = 3$, $S_0 = 20$, $u = 1.2$, $d = 1/u$. Consider the trading strategy with the target function $\max(0, S_{t_3} - 20)$. Determine the development of the portfolio valuc, and the number of stock units in the portfolio.

Let $F_k(S_{t_k})$ denote the value at $t_k$ of the portfolio that has the value $f(S_{t_n})$ at $t_n$. Here $f(s)$ is a given function, $f(s) = \max(0, s - K)$ for example. In the same way as before, we get

$$F_{k-1}(s) = qF_k(su) + (1-q)F_k(sd)$$

for $k = n, n-1, ..., 1$. Here $q = \frac{1-d}{u-d}$ and $F_n = f$.

If we solve these equations, we get

$$F_{n-1}(s) = qf(su) + (1-q)f(sd)$$

$$F_{n-2}(s) = qF_{n-1}(su) + (1-q)F_{n-1}(sd) =$$
$$q(qf(su^2) + (1-q)f(sud)) + (1-q)(qf(sdu) + (1-q)f(sd^2)) =$$
$$q^2 f(su^2) + 2q(1-q)f(sud) + (1-q)^2 f(sd^2).$$

..............

EXERCISE 3.11. Convince yourself that

$$F_{n-k}(s) = \sum_{j=0}^{k} \binom{k}{j} q^j (1-q)^{k-j} f(su^j d^{k-j}).$$

for $k = 0, 1, ..., n$.

The above result can also be written

$$F_{n-k}(s) = Ef(su^{X_k} d^{k-X_k}),$$

where $X_k$ is a random variable which is binomial distributed $(k, q)$. (Note that we have not made any assumption about the stochastic behavior of stock prices here. This is just a mathematical identity.) The number of stock units at $t_{k-1}$ equals

$$\frac{F_k(su) - F_k(sd)}{s(u-d)},$$

where $s = S_{t_{k-1}}$.

It is obvious that the target $f(S_{t_n})$ is related to the portfolio value at $t = 0$, but the following example can yet be instructive: Norm the stock price so that $S_0 = 1$, and choose $f(s) = e^s$. This portfolio ought to give a decent return if the stock price rises. In this case we have $F_0(1) > q^n f(u^n) = \exp(u^n - n \ln \frac{1}{q})$. If we norm the portfolio in such a way that the portfolio value equals 1 at $t = 0$, then the portfolio value at $t = t_n$ and the stock price $S_{t_n} = u^{n-k} d^k$ will satisfy

$$\frac{e^{u^{n-k}d^k}}{F_0(1)} < \exp\left(-u^n(1 - (\frac{d}{u})^k) + n \ln \frac{1}{q}\right) \to 0 \quad \text{when } n \to \infty \text{ if } k > 0.$$

On the other hand, it can be shown that the portfolio value at $t = t_n$ is of the order $q^{-n} \to \infty$ when $S_{t_n} = u^n$, but the risk is high if $n$ is large.

EXERCISE 3.12. Consider the case $n = 3$, $S_0 = 1$, $u = 1.2$, $d = 0.8$. Determine the portfolio development for the trading strategy that has the target function $\exp(s)$. Norm such that $F_0(1) = 1$.

### 3.2.2 An Asymptotic Result

Write $u = e^\delta$ and let $d = 1/u = e^{-\delta}$. In this case we have

$$F_{n-k}(s) = Ef(se^{Y_k}), \quad \text{where } Y_k = 2\delta(X_k - \frac{k}{2}).$$

We shall use this representation and the central limit theorem to approximate the portfolio value when $\delta$ is small and $k$ large.

EXERCISE 3.13. Show that

$$q = \frac{1}{1 + e^\delta} = \frac{1}{2}(1 - \frac{\delta}{2} + O(\delta^3)),$$

$$EY_k = 2\delta k(q - \frac{1}{2}) = -\frac{\delta^2 k}{2}(1 + O(\delta^2)), \text{ and}$$
$$\text{Var}(Y_k) = \delta^2 k 4q(1 - q) = \delta^2 k(1 + O(\delta^2))$$

as $\delta \to 0$.

It is seen here that $\delta^2 k$ must be of order 1 for the variance and expectation not to degenerate. Therefore assume that $\delta \to 0$ and $k \to \infty$ in such a way that $\delta^2 k \to v$, $0 < v < \infty$. It then follows from the central limit theorem that $Y_k$ converges in distribution to $Y_\infty$, where $Y_\infty$ is normally distributed with expectation $-\frac{v}{2}$ and variance $v$. Therefore also

$$Ef(se^{Y_k}) \to Ef(se^{Y_\infty})$$

if $f$ is regular.

It is seen in the following way what happens to the number of stock units: Put $s_1 = se^{-\delta}$, $s_2 = se^{\delta}$ and $G(x) = f(xe^{Y_k})$. We have

$$\frac{f(se^{\delta}e^{Y_k}) - f(se^{-\delta}e^{Y_k})}{se^{\delta} - se^{-\delta}} = \frac{G(s_2) - G(s_1)}{s_2 - s_1} = G'(s_3) = f'(s_3e^{Y_k})e^{Y_k}$$

for some $s_3$ between $s_1$ and $s_2$. Both $s_1$ and $s_2$ tend to $s$ as $\delta \to 0$ and hence the same holds for $s_3$. The expected value of the above expression therefore converges to

$$E[f'(se^{Y_\infty})e^{Y_\infty}] = \frac{d}{ds}Ef(se^{Y_\infty})$$

if $f$ is sufficiently regular.

Summing up:

**Theorem 3.2.** *Assume that $f$ is regular. Then*

$$Ef(se^{Y_k}) \to Ef(se^{-\frac{v}{2}+\sqrt{v}Z})$$

$$\frac{Ef(se^{\delta}e^{Y_k}) - Ef(se^{-\delta}e^{Y_k})}{se^{\delta} - se^{-\delta}} \to \frac{d}{ds}Ef(se^{-\frac{v}{2}+\sqrt{v}Z})$$

*as $\delta \to 0$ and $k \to \infty$ in such a way that $\delta^2 k \to v$, $0 < v < \infty$. Here $Z$ is normally distributed with expectation 0 and variance 1.*

It can be shown that sufficient regularity is that $f(s)$ is continuous for $s > 0$ and does not grow too fast: There are positive constants $C$ and $m$ such that $|f(s)| \leq C(s^m + s^{-m})$ for all $s > 0$. The proof is, however, fairly technical.

On the other hand, the above sketch of a proof can be made rigorous by the mathematically minded reader under the assumption that $f$ has a continuous derivative except at finitely many points and satisfies the Lipschitz condition $|f(s_1) - f(s_2)| \leq C|s_1 - s_2|$ for some constant $C$ and all $s_1 > 0$ and $s_2 > 0$. Note that the functions $f(s) = \max(0, s - K)$ and $f(s) = \max(0, K - s)$ satisfy this condition.

## 3.3 IMPLEMENTING TRADING STRATEGIES

The initial value of the stock, $S_0$, is irrelevant. The relevant thing is the value of $S_t/S_0$.

Therefore norm the stock price so that $S_0 = 1$, and start with a binary tree with $b_{0,0} = 1$, $b_{k+1,j} < b_{k,j} < b_{k+1,j+1}$. After that, the binary tree for the weights $q_{k,j}$ is calculated:

$$q_{k-1,j} = \frac{b_{k-1,j} - b_{k,j}}{b_{k,j+1} - b_{k,j}}.$$

This step is simplified if we have chosen $b_{k,j}$ such that $q_{k,j} = q$ independently of $k$ and $j$.

Then the binary tree for the portfolio values $F_{k,j}$, is calculated:

$$F_{k-1,j} = q_{k-1,j}F_{k,j+1} + (1 - q_{k-1,j})F_{k,j},$$

$k = n, n-1, ..., 1$, where $F_{n,j} = f(b_{n,j})$. If the initial capital is given, the portfolio value is normalized such that $F_{0,0}$ equals 1. This can be done by dividing $F_{k,j}$ by $F_{0,0}$. Alternatively, it may happen that $f$ depends on a few parameters, and these can be determined so that $F_{0,0} = 1$. We shall now assume that this has been done: $F_{0,0} = 1$.

After that, the tree for the number of stock units is calculated:

$$a_{k-1,j} = \frac{F_{k,j+1} - F_{k,j}}{b_{k,j+1} - b_{k,j}}.$$

Finally the cash holding is calculated:

$$c_{k,j} = F_{k,j} - a_{k,j}b_{k,j}.$$

Let $j_k$, $k = 0, ..., n$, denote the numbers that satisfy $S_{t_k} = b_{k,j_k}$

The portfolio value, $P_t$, at time $t$ is given by

$$P_t = P_0\big(c_{k,j_k} + a_{k,j_k}S_t\big) \text{ for } t_k \le t \le t_{k+1},\ k = 0, ..., n-1.$$

This, however, presupposes that we trade the stock exactly at the levels of the binary tree which hardly is possible in real life. Assume instead that we trade at $\tau_1, \tau_2, ...$, where hopefully $S_{\tau_k}$ is close to $b_{k,j_k}$. (In the example, Framfab ($\tau_k$ is the first moment $\ge t_k$ for which the stock is at or has passed this level of the binary tree.)

Let $Q_t$ denote the value of the portfolio which is rebalanced at these moments. Then $Q_t = P_t$ for $t \le t_1$. After that, the portfolios part and the value of $Q_t$ after $\tau_1$ depends on how the rebalancing is done, and this can be done in different ways.

Here is an example of how to proceed. The portfolio value at $t \le \tau_1$ equals

$$Q_t = P_0(c_{0,0} + a_{0,0}S_t).$$

After the rebalancing at $t = \tau_1$, one can try

$$P_0(c_{1,j_1} + a_{1,j_1}S_t),$$

but this value differs slightly from $Q_{\tau_1}$ when $t = \tau_1$. Correct this difference by multiplying by

$$\frac{Q_{\tau_1}}{P_0(c_{1,j_1} + a_{1,j_1}S_{\tau_1})}.$$

By continuing in this way, one gets

$$Q_t = K_k(c_{k,j_k} + a_{k,j_k} S_t) \quad \text{for } \tau_k \leq t \leq \tau_{k+1},\ k = 0, ..., n-1,$$

where

$$K_k = Q_{\tau_k} / (c_{k,j_k} + a_{k,j_k} S_{\tau_k}).$$

### 3.3.1 Portfolio Insurance

A way to reduce the risk of a stock holding is to choose a trading strategy with a target function that is never considerably below the initial value of the portfolio.

Here we shall let the initial value of the portfolio equal 1 and shall consider target functions of the form

$$f(s) = \begin{cases} g & \text{if } s \leq b \\ g + l(s-b) & \text{if } b < s \leq c \\ g + l(c-b) & \text{if } s > c \end{cases}$$

or shorter: $f(s) = g + l(s-b)_+ - l(s-c)_+$.

Here $0 \leq b < c \leq \infty$ are break points, whereas $g$, $0 \leq g < 1$, is a floor for the portfolio value. The parameter $l$ is the slope of the line between the breakpoints, and $g + l(c-b)$ is a ceiling for the portfolio value. The parameters have to be chosen so that the portfolio value at $t = 0$ equals 1.

Consider first the case $c = \infty$. If for a given breakpoint $b$ we reduce the risk by increasing $g$, we have to pay with a smaller $l$, and hence a lower portfolio value for stock values above the breakpoint $b$. The option strategy "Protective put" in Section 2.3.3 has a target function of this form.

Adding a ceiling to the target function at a breakpoint $c > b$ can be suitable when one expects only moderate rises of the stock price. The advantage is that for a given $g$, one can increase the slope $l$. Alternatively, raise the floor $g$ for a given slope. The target functions for the trading strategies "Finance the insurance with a short position in a call" and "Covered call" are of this form with $b = g$ and $l = 1$ and with $b = g = 0$, respectively.

One way to compute $F_{k,j}$ is: Choose $b$ and $c$, and define the function $h$ by

$$h(s) = \begin{cases} 0 & \text{if } s \leq b, \\ s-b & \text{if } b < s \leq c, \text{ and} \\ c-b & \text{if } s > c. \end{cases}$$

Then compute

$$H_{k-1,j} = q_{k-1,j} H_{k,j+1} + (1 - q_{k-1,j}) H_{k,j}$$

for $k = n, n-1, ..., 1$, where $H_{n,j} = h(b_{n,j})$. We then have the identities

$$F_{k,j} = g + lH_{k,j},$$

and hence $g$ and $l$ have to be chosen such that

$$g + lH_{0,0} = 1.$$

Which one of the stock and the insured portfolio that performs best will depend on which way the stock chooses. An essential difference is that the insured portfolio has lower risk.

***Example 3.3.*** In Fig. 3.3 the development of the stock Ericsson (unbroken) and the insured portfolio during the period July 21, 1998–November 1, 2001 are shown.

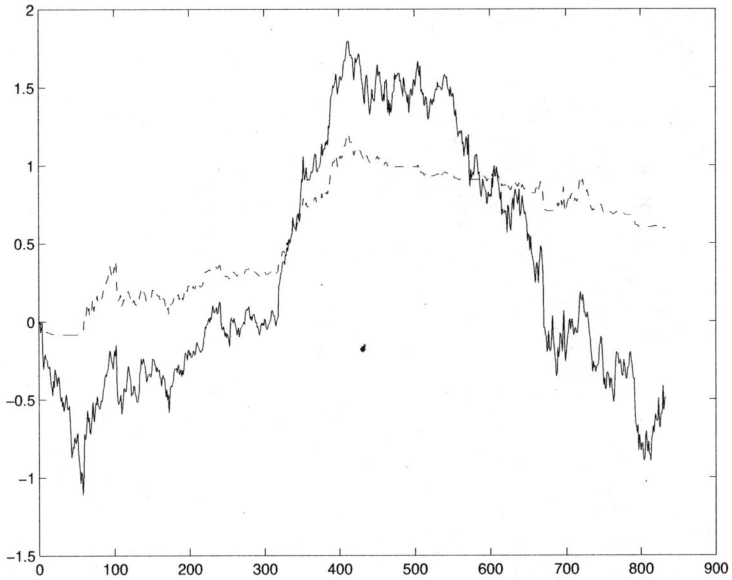


**Figure 3.3.** The development of Ericsson and the insured portfolio. Logarithmic scale.

The binary tree is given by the model of Cox, Ross, and Rubinstein with $d = 1/u$:

$$b_{k,j} = u^j d^{k-j} = u^{2j-k}, \quad k = 0, 1, ..., n, j = 0, ..., k$$

with the parameters $u = 1.1$ and $n = 8$. The trading strategy is given by $b = 1$, $c = \infty$, and $g = 0.95$. These give $l = 0.48$. The scale is logarithmic. Each unit on the vertical axis corresponds to a doubling/halving. The volatilities were $0.65$ and $0.21$ for the stock and the portfolio, respectively.

If the stock falls sufficiently (in this case $32\%$) during one period $t_0, ..., t_n$, then the entire value of the portfolio is in the cash account, and it will remain so until we start again at $t = t_n$. In this example, this happened a few times, around day 50 for example.

# CHAPTER 4

# STOCHASTIC PROPERTIES OF STOCK PRICES

The most frequently used stochastic model for stock prices is geometric Brownian motion. This model implies that the logarithmic returns are normally distributed. It is wellknown that this does not hold in reality for logarithmic returns over short time intervals. This will cause problems. It is, for example, not immediately clear how to define volatility.

We do not propose a parametric model that is more realistic than geometric Brownian motion. Instead we verify empirically that stock prices have certain properties that will imply that the volatility is well-defined, and also that some identities implied by the Gaussian assumption hold in reality. Some of these identities, but not all; the variance of the volatility estimator is unfortunately not as small as the Gaussian assumption implies.

For the remainder of the book, it is of utmost importance to be able to estimate the volatility in particular, and covariances of returns in general. They are needed to calculate the weights for the different trading strategies in the following chapters. We show that it is possible to do this, but that it is not possible to estimate the expected return or the drift (expected growth) of a single stock with sufficient precision.

In this chapter we also make clear the difference between expected return and drift. High drift is desirable. The expected return can take any value for a given drift, but it is the drift that determines the long-term development of an asset. This insight will have consequences for Chapter 6.

Here we also introduce the "*Five Stocks*", the developments of the stock index Affärs- världens generalindex (AFGX), and the stocks AstraZeneca (AZN), Ericsson B (LME), Hennes (HM), Skandia (SDIA), and Skanska (SKA) during the period October 7, 1996–October 31, 2000. In total 1025 trading days.

The daily closing prices for these are plotted in Fig. 4.1. The initial value is normed to 1. Note that two different scales have been used.

The data will be used throughout the book to illustrate different methods and portfolios.

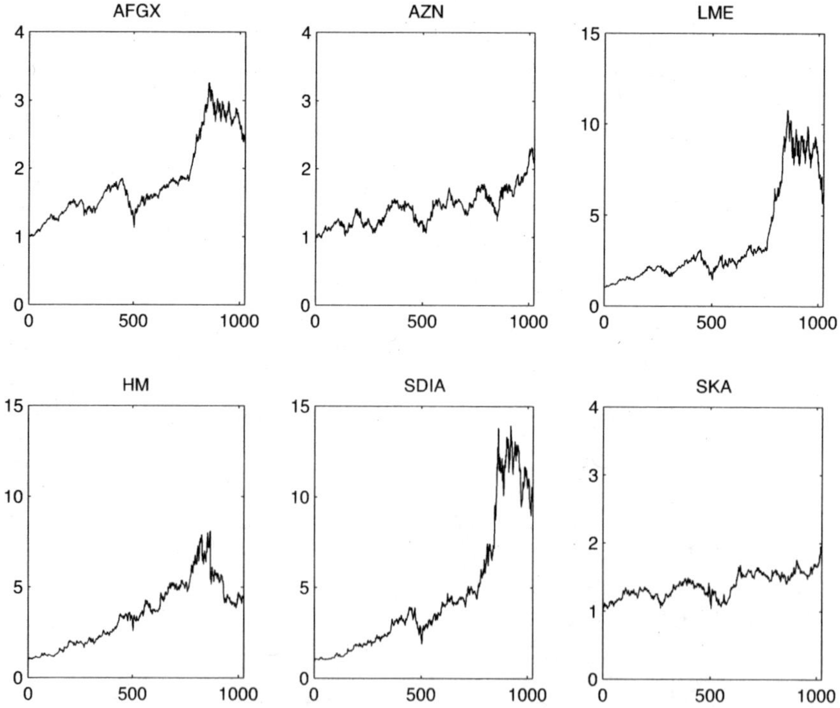


**Figure 4.1.** The development of *Five Stocks*.

## 4.1 GROWTH

Let $S(t)$ denote the price of an asset at time $t$. Unless otherwise stated, we shall measure $t$ in years. One year consists of about 250 trading days. The daily closing prices therefore are $S(0), S(\partial t), S(2\partial t), \ldots$, where $\partial t = 1/250$.

The *growth* in the interval $(t_1, t_2)$, $G(t_1, t_2)$ is defined as

$$G(t_1, t_2) = \ln \frac{S(t_2)}{S(t_1)}.$$

This can also be written

$$S(t_2) = S(t_1)e^{G(t_1,t_2)}.$$

We have

$$G(t_1, t_2) = G(t_1, t_3) + G(t_3, t_2)$$

for all $t_1, t_2, t_3$ and hence also

$$G(0, n\partial t) = \sum_{k=1}^{n} G((k-1)\partial t, k\partial t).$$

EXERCISE 4.1. Show this.

### 4.1.1 The Distribution of the Growth

We shall now compare the distribution of the growth with the normal distribution. In Fig. 4.2 a histogram over the daily growths for Ericsson B is shown together with the density of the normal distribution with the same expected value and variance as the observations.

It is seen here that the normal distribution underestimates the probability for small and large changes and overestimates the probability changes between them; roughly 1 to 2 standard deviations. The standard deviation equals $0.031$, and the mean value is $0.002$. The largest and smallest observations are $0.19$ and $-0.18$, respectively. This amounts to about 6 standard deviations. The corresponding normal probability equals $2(1 - \Phi(6)) \approx 2 \times 10^{-9}$ and is therefore entirely negligible. The daily growths are thus not normally distributed in this case. Also the others of the *Five Stocks* have the properties described above.

AFGX is a weighted sum, and one can therefore expect that this because of the central limit theorem is approximately normally distributed. Figure 4.3, however, shows that this is not the case.

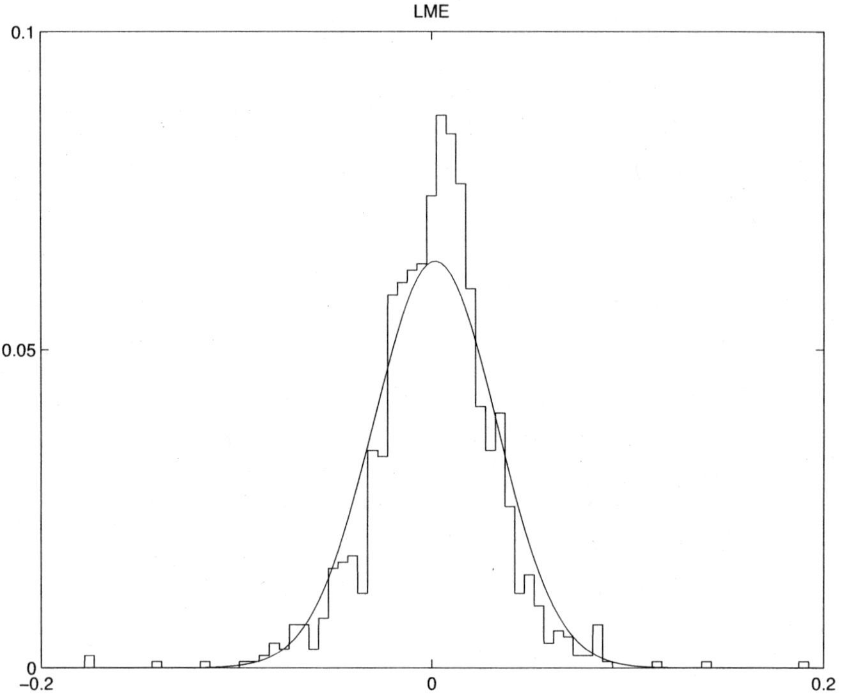


**Figure 4.2.** The daily growths of LME.

In this case the standard deviation equals $0.015$, and the mean value $0.001$. The largest and the smallest observations are $0.10$ and $-0.07$, respectively, about 7 and 5 standard deviations, respectively.

The above holds for the daily growths. In order to study the growth over periods longer than 1 day, we shall change over to normal plots because the number of observations will be too few to make informative histograms.

Assume that $X_1, ..., X_n$ are independent and normally distributed random variables that have the expected value $\nu$ and the standard deviation $\sigma$. Define

$$Y_k = \Phi(\frac{X_k - \nu}{\sigma}).$$

Then $Y_1, ..., Y_n$ are independent and uniformly distributed on the interval $(0, 1)$. Let $X_{(1)} < X_{(2)} < ... < X_{(n)}$ and $Y_{(1)} < Y_{(2)} < ... < Y_{(n)}$ be $X_1, ..., X_n$ and $Y_1, ..., Y_n$, respectively, ordered in increasing order.

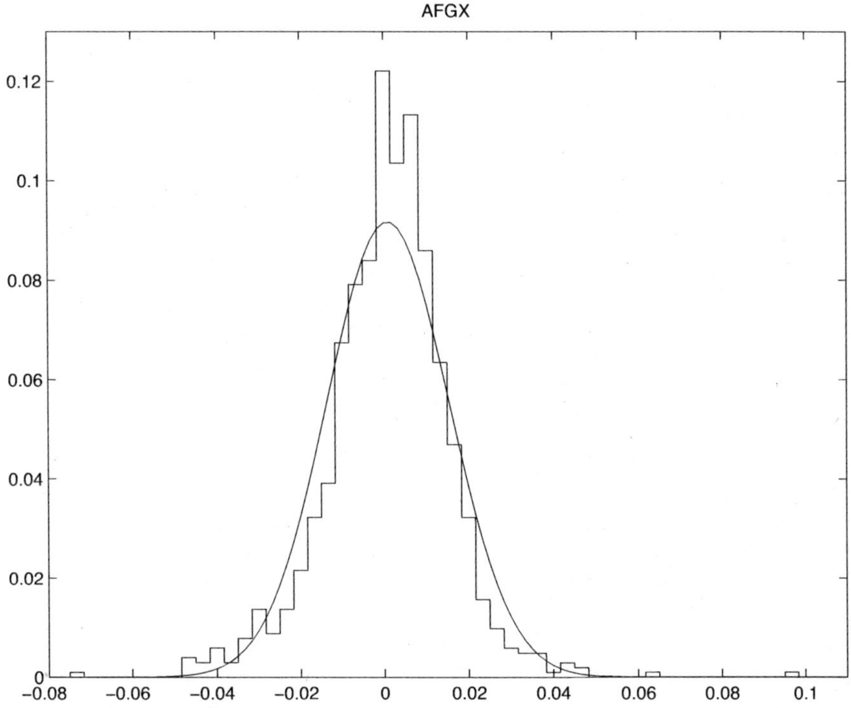


**Figure 4.3.** The daily growths of AFGX.

It can be shown that

$$EY_{(k)} = \frac{k}{n+1} \quad \text{and} \quad \text{Var}(Y_{(k)}) = \frac{k}{n+1}(1 - \frac{k}{n+1})\frac{1}{n+2}.$$

Therefore

$$Y_{(k)}) \approx \frac{k}{n+1}, \quad \text{i.e.,} \quad X_{(k)}) \approx \nu + \sigma z_{n,k},$$

where $\Phi(z_{n,k}) = \frac{k}{n+1}$.

If one plots $X_{(k)}$ against $z_{n,k}$ the points will therefore lie close to a straight line, provided that the normal distribution is at hand. In Fig. 4.4 and 4.5 the plots for *Five Stocks* are shown with $X_k = G((k-1)\partial t, k\partial t)$ for the cases $\partial t$ =1/250 (one day), 5/250 (one week), and 21/250 (one month).

All the plots of the daily growths have a distinct S-form. The weekly plots for the last three still have a slight S-form, but among the monthly plots only the one for Skandia possibly has. The growth over periods longer than about a month thus seems to be normally distributed.

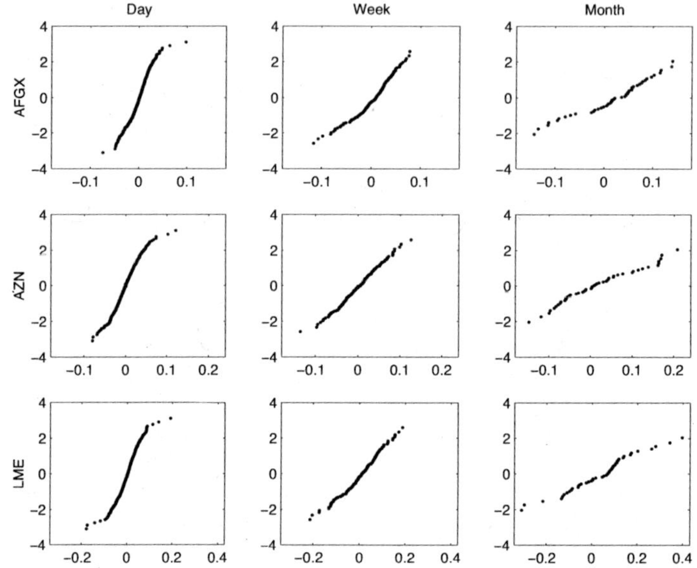


**Figure 4.4.**

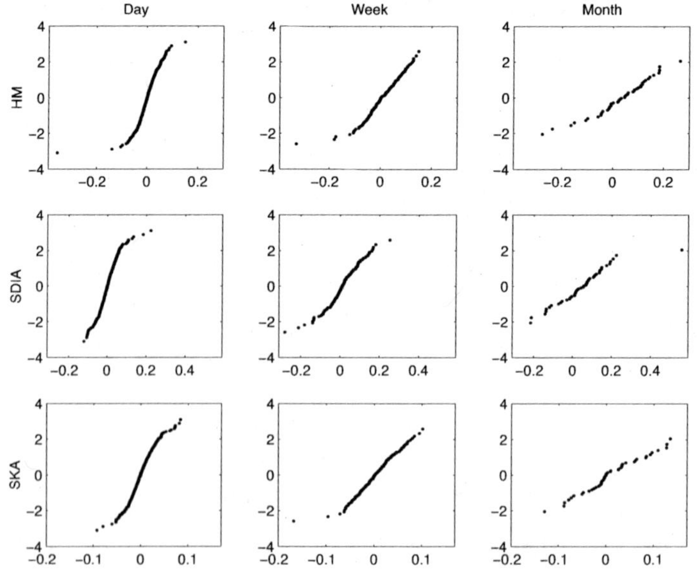


**Figure 4.5.** Normal plots for the growths of *Five Stocks*.

**Gaussian assumption.** *The growth over a period of length $T$ is approximately normally distributed, if $T$ is sufficiently large. ($T \geq 1/12$ will suffice.)*

Assume that we have observed $S(t)$ at the moments $0, \partial t, 2\partial t, ..., n\partial t = T$, where $\partial t = 1/250$ (one day). Let $X_k = G((k-1)\partial t, k\partial t)$ for $k = 1, ..., n$ denote the daily growths.

In order to investigate the dependence between the growths for two consecutive days, we shall calculate the correlations between $\boldsymbol{X}(0)$ and $\boldsymbol{X}(l)$ for $l = 0, 1, ..., m$. Here $0 < m < n$ and

$$\boldsymbol{X}(l) = (X_{1+l}, ..., X_{n-m+l}) \quad \text{for } l = 0, ..., m.$$

In Fig. 4.6 the correlation coefficients

$$\rho(l) = \frac{\text{Cov}(\boldsymbol{X}(0), \boldsymbol{X}(l))}{\sqrt{\text{Var}(\boldsymbol{X}(0))\text{Var}(\boldsymbol{X}(l))}}$$

are plotted as functions of $l$ for *Five Stocks* in the case $m = 20$. These correlations are dotted. As a comparison, also the correlations for the absolute values of the daily growths, $|X_k|$, are shown. The latter are marked by *.

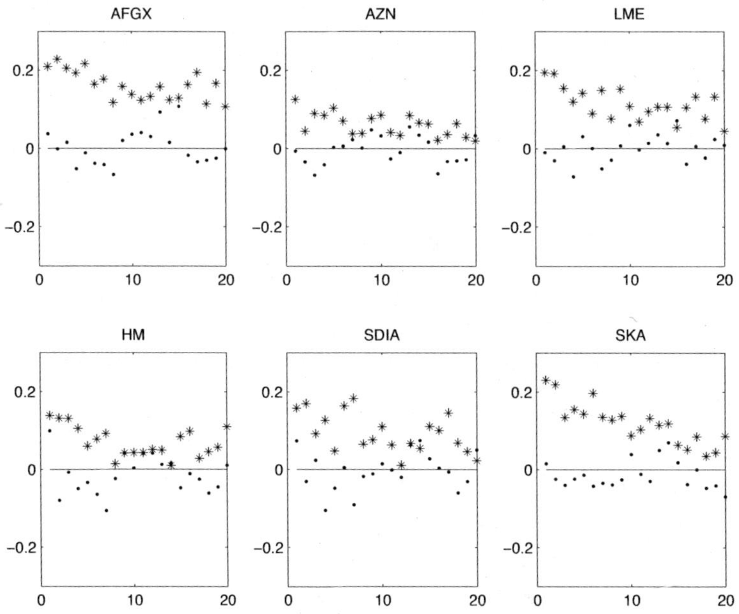


**Figure 4.6.** The correlations between the growths of the same asset two different days as a function of the distance between the days.

In Fig. 4.7 the moving averages

$$\bar{x}_k = (x_{k-5} + \cdots + x_{k+5})/11$$

are plotted for the first 95 days. It is seen that the growth on different days are uncorrelated, but they are not independent because the absolute values for adjacent days are positively correlated. The latter effect has died away after 90 days.

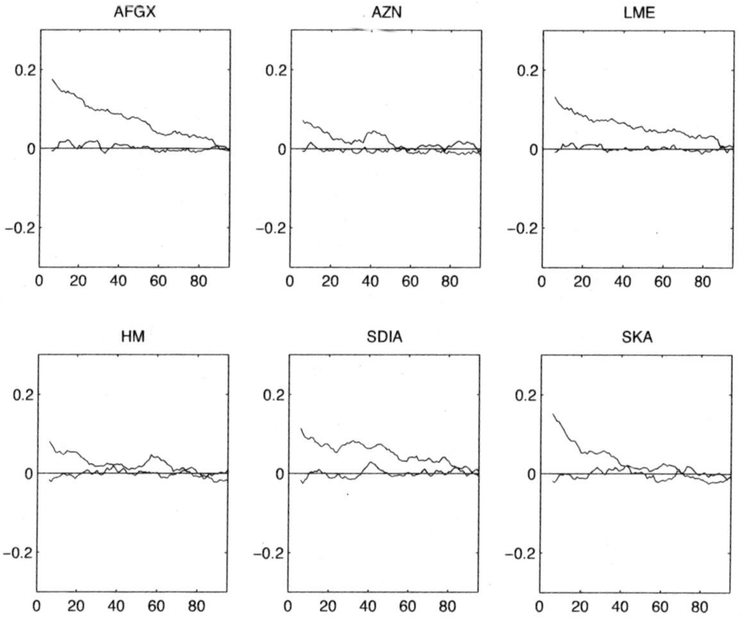


**Figure 4.7.** Moving averages of the correlations.

We shall make the following moment assumption:

**Moment assumption.** (a) *The expected value and variance for $G(0,t)$ exist and are continuous functions of $t$.*

(b) *For each pair $t_1 < t_2$ the expected value and the variance of $G(t_1+s, t_2+s)$ are the same for all $s$.*

(c) *The growths $G(t_1,t_2)$ and $G(t_2,t_3)$ are uncorrelated for all $t_1 < t_2 < t_3$.*

### 4.1.2 Drift and Volatility

Put $\nu(t) = EG(0,t)$ and $\sigma^2(t) = \text{Var}(G(0,t))$.

EXERCISE 4.2.

(a) Show that the moment assumption implies that

$$\nu(t+s) = \nu(t) + \nu(s) \text{ and } \sigma^2(t+s) = \sigma^2(t) + \sigma^2(s).$$

(b) Show that

$$\nu(t) = \nu(1)t \quad \text{and} \quad \sigma^2(t) = \sigma^2(1)t.$$

The parameter $\nu$ is called the *drift* and $\sigma$ the *volatility*.

We shall estimate the parameters $\nu$ and $\sigma$ by

$$\hat{\nu} = \bar{x}/\partial t \quad \text{and} \quad \hat{\sigma} = \sqrt{v(x)/\partial t}, \quad \text{respectively}$$

where

$$\bar{x} = \frac{x_1 + \cdots + x_n}{n} \quad \text{and} \quad v(x) = \frac{1}{n}\sum_{k=1}^{n}(x_k - \bar{x})^2.$$

The period October 7, 1996–October 31, 2000 was divided into 4 periods of equal length, 256 days. In Tables 4.1 and 4.2 the estimates of the volatility and the drift are based on data from the different periods are shown.

**Table 4.1.** Volatility

| Period | AFGX | AZN | LME | HM | SDIA | SKA |
|---|---|---|---|---|---|---|
| 1 | 0.15 | 0.29 | 0.31 | 0.34 | 0.35 | 0.22 |
| 2 | 0.27 | 0.33 | 0.54 | 0.41 | 0.56 | 0.32 |
| 3 | 0.19 | 0.30 | 0.48 | 0.32 | 0.37 | 0.27 |
| 4 | 0.28 | 0.35 | 0.60 | 0.58 | 0.59 | 0.26 |
| 1–4 | 0.23 | 0.32 | 0.49 | 0.42 | 0.48 | 0.27 |

**Table 4.2.** Drift

| Period | AFGX | AZN | LME | HM | SDIA | SKA |
|---|---|---|---|---|---|---|
| 1 | 0.39 | 0.15 | 0.75 | 0.63 | 0.68 | 0.15 |
| 2 | −0.11 | −0.05 | −0.13 | 0.49 | 0.27 | 0.05 |
| 3 | 0.39 | 0.40 | 0.74 | 0.49 | 0.65 | 0.19 |
| 4 | 0.22 | 0.23 | 0.43 | −0.15 | 0.60 | 0.27 |
| 1–4 | 0.22 | 0.18 | 0.45 | 0.36 | 0.55 | 0.17 |

Only the first and last observation are needed to calculate $\hat{\nu}$:

EXERCISE 4.3. Show that

$$\hat{\nu} = \frac{\ln(S_T/S_0)}{T}.$$

A consequence of the Gaussian assumption is that

$$\frac{\hat{\nu} - \nu}{\sigma/\sqrt{T}}$$

is approximately normally distributed with expectation 0 and variance 1. We shall see below that $\hat{\sigma}$ is a consistent estimator of $\sigma$. Therefore also

$$\frac{\hat{\nu} - \nu}{\hat{\sigma}/\sqrt{T}}$$

is approximately standardized normally distributed.

Therefore, for $T$ large

$$\hat{\nu} \pm \frac{\hat{\sigma}}{\sqrt{T}} z_{\epsilon/2}$$

is a confidence interval for $\nu$ with the approximate confidence level $1 - \epsilon$. Here $\Phi(z_p) = 1 - p$.

EXERCISE 4.4.

(a) Determine an approximate 95% confidence interval for the drift of Ericsson based on data from Period 1– 4.

(b) Roughly, how long a period of observations would be needed in order to give the length 5% to the 95% confidence interval?

If one lowers the confidence level to 90% and increases the interval length to 10%, then the length of the observation period decreases to $\approx 250$ years. With the confidence level 50% and interval length 10%, the length of the observation period becomes $\approx 50$ years.

**Conclusion 4.1.** *It is not possible to estimate the drift of a stock in this way with any reasonable precision.*

The volatilities vary with the period. Thus for example Period 2 is more volatile than Period 1. This makes it more difficult to use historical data to estimate the future volatility. We shall, however, see that what is needed is the relative volatilities. In Table 4.3 the ratios

$$100\sigma_j/(\sigma_1 + \cdots + \sigma_6), \quad j = 1, ..., 6$$

are calculated for the different periods. The figures are rounded to the nearest integer.

It is thus possible to estimate the relative volatilities from historical data.

**Table 4.3.** Relative volatilities

| Period | AFGX | AZN | LME | HM | SDIA | SKA |
|---|---|---|---|---|---|---|
| 1 | 9 | 17 | 19 | 20 | 21 | 13 |
| 2 | 11 | 14 | 22 | 17 | 23 | 13 |
| 3 | 10 | 16 | 25 | 17 | 19 | 14 |
| 4 | 11 | 13 | 23 | 22 | 22 | 10 |
| 1– 4 | 10 | 15 | 22 | 19 | 22 | 12 |

### 4.1.3 The Stability of the Volatility Estimator

Here we shall study the relation between the period length $\partial t$ and the stability of the volatility estimator.

Note that

$$\hat{\sigma}^2 = \frac{1}{n\partial t}\sum_{k=1}^{n} x_k^2 - \partial t\hat{\nu}^2.$$

The expressions $\partial t\hat{\nu}^2$ based on data from Period 1– 4 equal $0.0002, 0.0001, 0.0008,$ $0.0005, 0.0012$, and $0.0001$, respectively. These are negligible compared to $\hat{\sigma}^2$. We can therefore equally well use the estimator

$$\breve{\sigma}^2 = \frac{1}{n\partial t}\sum_{k=1}^{n} x_k^2$$

for $\sigma^2$ instead of $\hat{\sigma}^2$.

We have seen that the growths over sufficiently long periods are approximately normally distributed.

EXERCISE 4.5. Assume that the growths are independent and normally distributed. In this case we thus have that

$$\frac{n\hat{\sigma}^2}{\sigma^2}$$

is $\chi^2$-distributed with $n-1$ degrees of freedom and is independent of $\hat{\nu}$.

(a) Show that

$$E\hat{\sigma}^2 = \frac{n-1}{n}\sigma^2 \quad \text{and} \quad \text{Var}(\hat{\sigma}^2) = \frac{n-1}{n}\frac{2\sigma^4}{n}.$$

(b) Show that $\hat{\sigma}$ is asymptotically normally distributed with expectation $\sigma$ and variance $\frac{\sigma^2}{2n}$), and hence

$$\hat{\sigma}(1 \pm \frac{z_{\epsilon/2}}{\sqrt{2n}})$$

is a confidence interval for $\sigma$ with the approximate confidence level $1-\epsilon$.

Under the above assumptions, the length of the confidence interval is thus proportional to

$$\sqrt{\frac{1}{n}} = \sqrt{\frac{\partial t}{T}}.$$

The precision of the estimate therefore improves if we make more frequent observations, that is, decrease $\partial t$. This does not hold for the estimate of $\nu$.

In the real world this seems to hold only to a certain limit. We shall use $\breve{\sigma}^2$ instead of $\hat{\sigma}^2$ and will assume that the covariance between $X_i^2$ and $X_j^2$ only depends on $\partial t$ and the distance $|j - i|$. Under these assumptions, we have

$$\text{Var}(\breve{\sigma}^2) = \frac{1}{(n\partial t)^2} \sum_{i=1}^{n} \sum_{j=1}^{n} \text{Cov}(X_i^2, X_j^2) = \frac{1}{n\partial t^2} \text{Var}(X_1^2)(1 + 2R_n).$$

Here

$$R_m = \sum_{k=1}^{m} \rho_k (1 - \frac{k}{n}),$$

where $\rho_k$ stands for the correlation coefficient between $X_1^2$ and $X_{1+k}^2$.

EXERCISE 4.6. Show this.

If one estimates these correlation coefficients with data from Period 1– 4, one finds that they have roughly the same general behavior as the correlation coefficients for the absolute values but die away somewhat faster. The sequence $R_1, R_2, ...$ increases to begin with and then levels off at $m\partial t \approx 60$ days.

In Fig. 4.8, $\text{Var}(\breve{\sigma}^2)$ is plotted for $\partial t = 1, 2, ..., 30$ days. The three lines are

$$2\sigma^4/n = 2\sigma^4 \partial t/1024$$

with the estimate of $\sigma$ based on daily ($\partial t = 1$), weekly ($\partial t = 5$), and bi-weekly data ($\partial t = 10$), respectively. Note that we have used two different scales.

Disregarding Skandia, it is maybe possible to to approximate the point clusters with a straight line for $\partial t \geq 7$, but not for $\partial t \leq 3$. In particular, $\partial t = 2$ gives high variance. The variances attain minimum for $\partial t$ =7, 10, 9, 16, 7, and 8 days, respectively. After that, the curves straighten out. This deviation from the line can be explained by the growth not being normally distributed when $\partial t$ is small.

Hence there seems to be a limit for $\partial t$ below which the variances do not decrease.

In Figs. 4.9 and 4.10, normal plots for $\breve{\sigma}^2$ are shown for the cases $\partial t = 1$ day, and $n = 1, 63, 125$ (day, quarter, and half year, respectively). In this case a period longer than one month is required in order to approximate the distribution by the normal distribution, but also in order to get a stable estimate of the volatility. (See Exercise 4.9 below.)

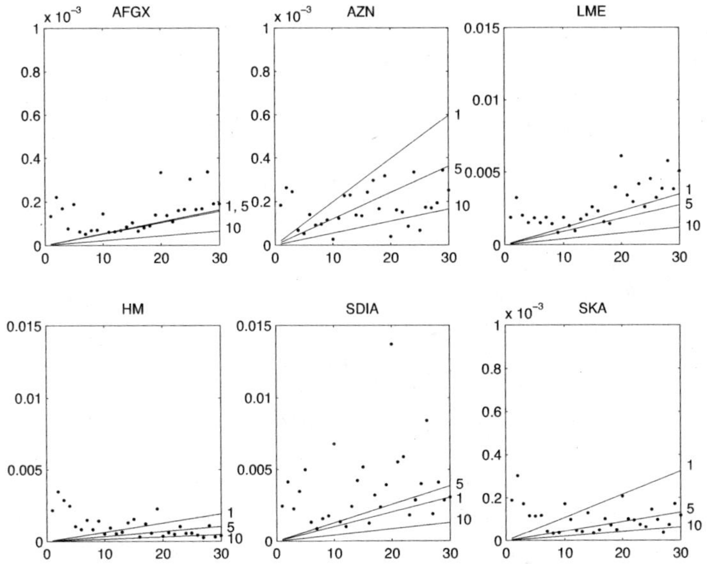


**Figure 4.8.** The variance of the estimate of the variance for different period lengths.

Note that the $35\%$ decrease of the growth for HM day 870 is still seen on the half-year level.

EXERCISE 4.7. Assume that

$$\breve{\sigma}^2 = \sigma^2 + \mathrm{SD}(\breve{\sigma}^2)Z,$$

where $Z$ is approximately normally distributed $(0, 1)$.

(a) Show that

$$\breve{\sigma} = \sigma\Big(1 + dZ + O(d^2)\Big),$$

as $d \to 0$. Here $d = \frac{\mathrm{SD}(\breve{\sigma}^2)}{2\sigma^2}$.

(b) Show that

$$\breve{\sigma}(1 \pm z_{\epsilon/2}\hat{d})$$

is a confidence interval for $\sigma$ with the approximate confidence level $1 - \epsilon$, provided $\hat{d}$ is small. Here $\hat{d}$ is a consistent estimator of $d$.

The values of $\hat{d}$ and $\breve{\sigma}$ shown in Tables 4.4 and 4.5 were obtained from data from Period 1–4.

There is no indication that any of the period lengths is decidedly best. However, it is seen that the formula $2\sigma^4\partial t/T$ does not describes $\mathrm{Var}(\breve{\sigma}^2)$ very well.

EXERCISE 4.8. Show that if $\mathrm{Var}(\breve{\sigma}^2) = 2\sigma^4\partial t/T$, then $d = \sqrt{\frac{\partial t}{2T}}$.

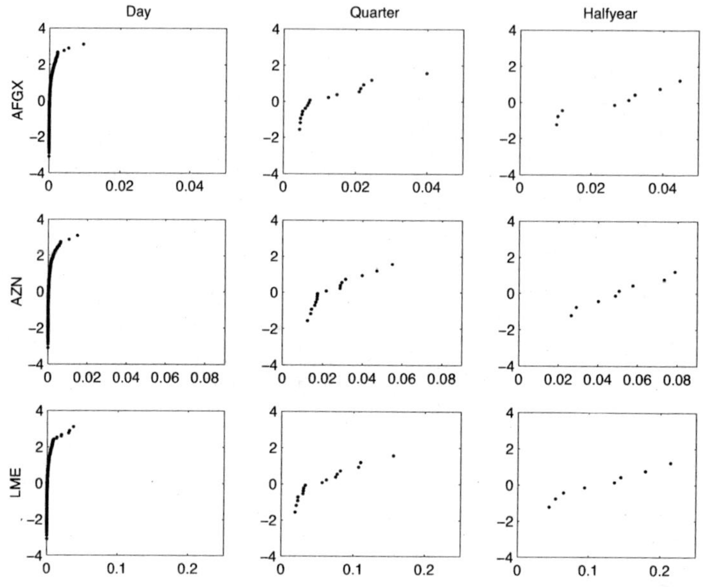


**Figure 4.9.**

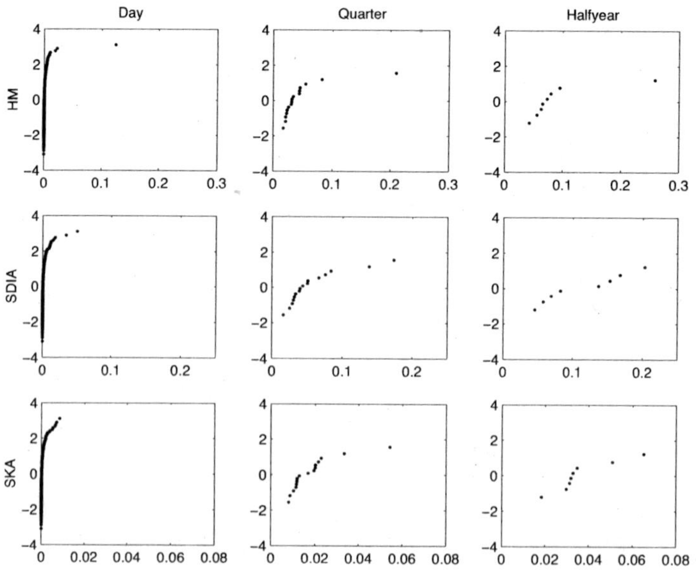


**Figure 4.10.** Normal plots for the estimates of the variance.

**Table 4.4.** Values of $\hat{d}$

| $\partial t$ | AFGX | AZN | LME | HM | SDIA | SKA |
|---|---|---|---|---|---|---|
| 1 | 0.11 | 0.07 | 0.09 | 0.13 | 0.11 | 0.09 |
| 5 | 0.12 | 0.04 | 0.09 | 0.09 | 0.15 | 0.09 |
| 10 | 0.11 | 0.03 | 0.10 | 0.08 | 0.16 | 0.11 |

**Table 4.5.** Values of $\breve{\sigma}$

| $\partial t$ | AFGX | AZN | LME | HM | SDIA | SKA |
|---|---|---|---|---|---|---|
| 1 | 0.23 | 0.32 | 0.49 | 0.43 | 0.48 | 0.27 |
| 5 | 0.24 | 0.29 | 0.49 | 0.43 | 0.49 | 0.25 |
| 10 | 0.24 | 0.27 | 0.46 | 0.39 | 0.51 | 0.25 |

This gives $d = 0.022, 0.049, 0.070$ for $\partial t = 1, 5$, and 10 days, respectively. These values are too optimistic, at least for $\partial t = 1$ and 5.

In the above example the length of the observation period was 1024 days, which is $T = 1024/250 = 4.096$ years. We shall now give the dependence of the confidence interval on $T$. For long periods of time, $\breve{\sigma}$ is approximately normally distributed and the correlation between disjoint and long periods of time should be negligible. Therefore

$$\breve{\sigma}(1 \pm z_{\epsilon/2}\frac{\hat{c}}{\sqrt{T}})$$

is a confidence interval for $\sigma$ with the approximate confidence level $1 - \epsilon$. Here

$$c = \frac{\sqrt{T}\mathrm{SD}(\breve{\sigma}^2)}{2\sigma^2}.$$

The relation between $d$ in the above example and $c$ thus is $c = \sqrt{1024/250}d$ and a typical value is therefore given by $c = 0.2$.

EXERCISE 4.9. Assume that $c = 0.2$.

(a) Roughly, how large does $T$ need to be in order for $\breve{\sigma}(1 \pm 0.1)$ to be a confidence interval with the confidence levels 95%, 75%, and 50%, respectively?

(b) The same question for the interval $\breve{\sigma}(1 \pm 0.2)$.

*Information:* $z_{0.25} = 0.67$, $z_{0.125} = 1.15$, $z_{0.025} = 1.96$.

We thus cannot measure the volatility with the same high precision we are perhaps used to from textbooks in statistics, but the situation is not as hopeless as for the drift. We can get a decent understanding of the mutual relations of the volatilities.

EXERCISE 4.10. Use the daily growths during Period 1–4 to determine confidence intervals for $\sigma$ with the approximate confidence levels 90% and 50% for one of the stocks.

It can be seen from the answer to this exercise, for example, that $\sigma_{afgx} < \sigma_{azn} < \sigma_{sdia}$ and that we cannot decide which of the stocks LME and SDIA has the highest volatility.

**Summary 4.2.** *The estimate of the volatility is approximately normally distributed as $T$ is large. The standard deviation of the estimate has the form $c(\partial t)/\sqrt{T}$. There is a limit for $\partial t \approx 10$ days, below which $c(\partial t)$ does not decrease. In particular, the identities in Exercise 4.8 do not hold for small $\partial t$.*

## 4.2 RETURN

The *return* of an asset in the time interval $(t_1, t_2)$, $R(t_1, t_2)$ is defined by

$$R(t_1, t_2) = \frac{S(t_2) - S(t_1)}{S(t_1)}.$$

The relation between return and growth is:

$$R(t_1, t_2) = e^{G(t_1, t_2)} - 1, \quad \text{or} \quad G(t_1, t_2) = \ln\big(1 + R(t_1, t_2)\big).$$

EXERCISE 4.11. Show that

$$R(t_1, t_2) = \Big(1 + R(t_1, t_3)\Big)\Big(1 + R(t_3, t_2)\Big) - 1.$$

We shall relate the expected value and the standard deviation of the return to the corresponding quantities for the growth — that is, to the drift and volatility.

EXERCISE 4.12. Assume that

$$G(0, \partial t) = \nu \partial t + \sigma Z \sqrt{\partial t},$$

where $Z$ is normally distributed $(0, 1)$. Show that

$$ER(0, \partial t) = (\nu + \frac{1}{2}\sigma^2)\partial t + O(\partial t^2) \text{ and } \mathrm{Var}\Big(R(0, \partial t)\Big) = \sigma^2 \partial t + O(\partial t^2).$$

Define

$$\mu = E[R(0, \partial t)]/\partial t \text{ and } \sigma_r^2 = \mathrm{Var}\big(R(0, \partial t)\big)/\partial t,$$

where $\partial t = 1/250 =$ one day.

Let

$$r_k = r_k(\partial t) = R((k-1)\partial t, k\partial t), \quad k = 1, ..., n.$$

The expected value $\mu$ and the variance $\sigma_r^2$ are estimated by

$$\hat{\mu} = \frac{\bar{r}}{\partial t} = \frac{r_1 + \cdots + r_n}{n\partial t} \text{ and } \hat{\sigma}_r^2 = \frac{v(r)}{\partial t} = \frac{\sum_{k=1}^n (r_i - \bar{r})^2}{n\partial t},$$

respectively.

The estimated values based on data from Period 1–4 are shown in Table 4.6.

**Table 4.6.** Expected return, drift and volatility

| | AFGX | AZN | LME | HM | SDIA | SKA |
|---|---|---|---|---|---|---|
| $\hat{\sigma}_r$ | 0.2295 | 0.3191 | 0.4938 | 0.4163 | 0.4847 | 0.2735 |
| $\hat{\sigma}_r - \hat{\sigma}$ | 0.0002 | 0.0016 | 0.0001 | −0.0089 | 0.0064 | 0.0009 |
| $\hat{\mu}$ | 0.2489 | 0.2348 | 0.5706 | 0.4535 | 0.6654 | 0.2047 |
| $\hat{\mu} - \hat{\nu} - \frac{1}{2}\hat{\sigma}^2$ | 0.0001 | 0.0002 | 0.0003 | −0.0012 | 0.0015 | 0.0001 |

EXERCISE 4.13. Use normal approximation and data from Period 1–4 to calculate confidence intervals for the returns of the six assets with the approximate confidence level 90%.

In spite of the Gaussian assumption not holding for $\partial t = 1$ day there is hence a good agreement between $\sigma_r$ and $\sigma$, and between $\mu$ and $\nu + \frac{1}{2}\sigma^2$. It is also seen that $\sigma_r$ is so big in relation to $\mu$ that the estimate of the expected return suffers from the same uncertainty as the estimate of the drift.

**Summary 4.3.** *It is not possible to estimate $\mu$ in this way with any reasonable precision. The relations*

$$\mu = \nu + \frac{1}{2}\sigma^2 \quad \text{and} \quad \sigma_r = \sigma$$

*hold.*

If $\mu$ and $\nu$ are related as in the summary, and $-\frac{\sigma^2}{2} < \nu < 0$, then $\mu > 0$ but $\nu < 0$. We shall conclude this section by showing that it is the sign of the drift, $\nu$, and not the expected return, $\mu$, that decides the long-term development of an asset.

EXERCISE 4.14. Show that the moment assumption in Section 4.2 implies that $\text{Var}\Big(G(0,T)/T\Big) \to 0$ and hence also $G(0,T)/T \to \nu$ in probability as $T \to \infty$.

We have $S(T) = S(0)e^{G(0,T)}$. It therefore follows that $S(T) \to 0$ if $\nu < 0$ and $S(T) \to \infty$ if $\nu > 0$. If the yearly returns are stochastically independent and $\mu > 0$, then $E[S(T)] = S(0)(1+\mu)^T \to \infty$, which thus does not prevent that $S(T) \to 0$.

EXERCISE 4.15. Consider the following game: You get back twice the stake or a third of the stake with equal probabilities, 1/2. You play repeatedly, and the different games are independent. Your initial capital equals $K_0 = 1$. Let $K_n$ denote your capital after $n$ plays.

(a) Assume that you stake your entire capital each time. Calculate the expected return and the expected growth for each play. Show that $EK_n \to \infty$, but $K_n \to 0$ in probability as $n \to \infty$.

(b) Assume that each time you stake the proportion $p$ of your capital, $0 \leq p \leq 1$. Which values of $p$ maximize the expected return and the expected growth for each play? Show that $K_n \to \infty$ for the latter value.

## 4.3 COVARIATION

When constructing portfolios, it is of decisive importance to know the covariation of the assets measured by the correlation between the growths (or returns) of the assets.

Let $G_i(t_1, t_2)$ and $R_i(t_1, t_2)$, $i = 1, 2$, denote the growths and returns for two assets in the time interval $(t_1, t_2)$.

We have seen that the growths over disjoint time intervals for *one* stock are uncorrelated. There is therefore no reason to believe that anything different hold for the growths of two different stocks. Therefore

$$\text{Cov}\big(G_1((k-1)\partial t, k\partial t), G_2((k-1)\partial t, k\partial t)\big) =$$

$$\text{Cov}\big(G_1(0, \partial t), G_2(0, \partial t)\big) = \sigma_{1,2}\partial t.$$

Put

$$\sigma^r_{i,j} = \text{Cov}\big(R_i(0, \partial t), R_j(0, \partial t)\big)/\partial t$$

Let $r_{1,k}$, $r_{2,k}$ and $g_{1,k}$, $g_{2,k}$ denote observed returns and growths for the two assets in the time interval $((k-1)\partial t, k\partial t)$, where $\partial t$ is small. Assume that we have observed these during the time interval $[0, T]$ and that $n\partial t = T$.

The covariances $\sigma_{1,2}$ and $\sigma^r_{1,2}$ are estimated by

$$\hat{\sigma}_{1,2} = \frac{\sum_{k=1}^n (g_{1,k} - \bar{g_1})(g_{2,k} - \bar{g_2})}{n\partial t} \quad \text{and} \quad \hat{\sigma}^r_{1,2} = \frac{\sum_{k=1}^n (r_{1,k} - \bar{r_1})(r_{2,k} - \bar{r_2})}{n\partial t},$$

respectively.

The covariances $\sigma_{i,j}$ and $\sigma^r_{i,j}$ for *Five Stocks* were estimated from the daily growths and returns during Period 1– 4. The result is shown in Tables 4.7 and 4.8.

We shall therefore assume that $\sigma^r_{1,2}\partial t \approx \sigma_{1,2}\partial\partial t$ for $\partial t$ small.

The correlation coefficient,

$$\rho^r_{1,2} = \frac{\text{Cov}(r_1, r_2)}{\sqrt{\text{Var}(r_1)\text{Var}(r_2)}},$$

is therefore estimated by $\hat{\rho} = \hat{\sigma}_{1,2}/(\hat{\sigma}_1\hat{\sigma}_2)$, where $\hat{\sigma}_1$ and $\hat{\sigma}_2$ are the estimates of the volatilities.

**Table 4.7.** $\hat{\sigma}_{i,j}$

| | AFGX | AZN | LME | HM | SDIA | SKA |
|---|---|---|---|---|---|---|
| AFGX | 0.053 | 0.023 | 0.077 | 0.039 | 0.056 | 0.022 |
| AZN | | 0.101 | 0.031 | 0.026 | 0.043 | 0.022 |
| LME | | | 0.244 | 0.055 | 0.095 | 0.040 |
| HM | | | | 0.181 | 0.070 | 0.033 |
| SDIA | | | | | 0.229 | 0.043 |
| SKA | | | | | | 0.74 |

**Table 4.8.** $\hat{\sigma}^{r}_{i,j} - \hat{\sigma}_{i,j}$

| | AFGX | AZN | LME | HM | SDIA | SKA |
|---|---|---|---|---|---|---|
| AFGX | 0.0001 | 0.0002 | 0.0004 | −0.0002 | 0.0009 | 0.0001 |
| AZN | | 0.0010 | 0.0003 | 0.0004 | 0.0011 | 0.0002 |
| LME | | | 0.001 | 0.0012 | 0.0029 | 0.0006 |
| HM | | | | −0.0075 | 0.0014 | −0.0001 |
| SDIA | | | | | 0.0061 | 0.0004 |
| SKA | | | | | | 0.0005 |

**Table 4.9.** $\hat{\rho}_{i,j}$

| | AFGX | AZN | LME | HM | SDIA | SKA |
|---|---|---|---|---|---|---|
| AFGX | 1 | 0.31 | 0.68 | 0.40 | 0.51 | 0.35 |
| AZN | | 1 | 0.20 | 0.19 | 0.28 | 0.26 |
| LME | | | 1 | 0.26 | 0.40 | 0.30 |
| HM | | | | 1 | 0.34 | 0.28 |
| SDIA | | | | | 1 | 0.33 |
| SKA | | | | | | 1 |

The estimates of the correlation coefficients for *Five Stocks* based on data from Period 1–4 are given in Table 4.9.

It is seen, for example, that AZN has the lowest correlation with AFGX and that LME has the highest. If one divides Period 1–4 into the four sub periods Period 1,..., Period 4, the correlations between AZN and AFGX become $0.49$, $0.36$, $0.53$, $0.07$, respectively. The low correlation during Period 4 reflects the fact that pharmaceutical stocks are considered a safe haven during troubled times. During this period, AZN was slightly negatively correlated with LME, $-0.07$. On the contrary, the correlation between LME and AFGX was consistently high during the four periods: $0.66$, $0.61$, $0.82$, $0.67$, respectively. This can perhaps be explained by the weight of LME in the index being large at that time. The reason SDIA having high correlation with

AFGX is presumably different: The valuation of Skandia depended on the general development of the stock market.

In the following chapters we shall consider portfolios consisting of $m$ assets the prices of which at time $t$ are $S_1(t), ..., S_m(t)$. We have thus found that the following model agrees with reality quite well.

**Model A.**

$$S_j(t) = S_j(0)e^{\nu_j t + X_j(t)}. \tag{4.1}$$

*The process $\boldsymbol{X}(t) = (X_1(t), ..., X_m(t))$ has uncorrelated increments, expectation 0 and a covariance matrix of the form*

$$\text{Var}(\boldsymbol{X}(t)) = Qt.$$

We shall write $\sigma_{i,j}$ for the elements of $Q$.

We have also found that the model below holds approximately when $t$ is large.

**Model B.**

$$As\ Model\ A\ with\ \boldsymbol{X}(t)\ normally\ distributed. \tag{4.2}$$

In some cases this model gives results that agree with reality well even when $t$ is small. See Exercise 4.12 and Table 4.6. It is also fairly easy to make calculations using this model. For these reasons we shall on some occasions use it to try to find general relations.

In the case $m = 1$, Model B is called the *Black–Scholes model* or *geometric Brownian motion*.

A process $\{W_t,\ t \geq 0\}$ is a *Brownian motion*, or a *Wiener process* if it has independent increments and if $W_t$ is normally distributed with zero expectation and variance $t$.

It follows that $W_0 = 0$ and that $W_t - W_s$ is normally distributed with zero expectation and variance $|t - s|$. The latter can be seen by calculating the moment-generating functions of both sides of the identity $W_t = W_s + (W_t - W_s)$ when $t > s$.

The Black–Scholes model, or Model B in the case $m = 1$, can thus be written

$$S_t = S_0 e^{\nu t + \sigma W_t}, \tag{4.3}$$

where $W_t$ is a Wiener process.

Black–Scholes model implies that the increments of the stock price are as in the next exercise.

EXERCISE 4.16. Put $\partial S_t = S_{t+\partial t} - S_t$, where $\partial t$ is a small number, and let $E_t$ stand for the conditional expectation given $S_t$. Show that

$$E_t \partial S_t = S_t(\nu + \frac{\sigma^2}{2})\partial t + O((\partial t)^2), \ \ E_t(\partial S_t)^2 = S_t^2 \sigma^2 \partial t + O((\partial t)^2)$$

and

$$E_t(\partial S_t)^k = \begin{cases} O(\sqrt{\partial t}^k) & \text{for } k \text{ even,} \\ O(\sqrt{\partial t}^{k+1}) & \text{for } k \text{ odd.} \end{cases}$$

Conclude in particular that $\partial S_t$ is of the order of magnitude $\sqrt{\partial t}$, whereas $(\partial S_t)^2$ is of the same order of magnitude as $\partial t$.

*Hint:* Put $Z = \frac{\partial W_t}{\sqrt{\partial t}}$, and express $\partial S_t$ in terms of $Z$.

### 4.3.1 The Asymptotic Distribution of the Estimated Covariance Matrix

We shall later need the following result in order to determine the asymptotic distribution of the weights for certain portfolios.

**Theorem 4.4.** *Assume that the stock prices develop according to Model B. Let $\boldsymbol{Z}_n = \sqrt{n}(\hat{Q} - Q)$. Then $\boldsymbol{Z}_n$ converges in distribution to a normally distributed random matrix, $\boldsymbol{Z}$, having zero expectation, and*

$$E[Z_{i,k}Z_{j,l}] = \sigma_{i,j}\sigma_{k,l} + \sigma_{i,l}\sigma_{k,j}.$$

*Proof.* Let $\nu_p$ stand for the drift of asset $p$, $p = 1, ..., m$, and put $X_{p,h} = g_{p,h} - \nu_p$ for $h = 1, ..., n$. Then $EX_{p,h} = 0$ and

$$\hat{Q}_{i,k} = \frac{1}{n}\sum_{h=1}^{n}(X_{i,h} - \bar{X}_{i,\cdot})(X_{k,h} - \bar{X}_{k,\cdot}) = \frac{1}{n}\sum_{h=1}^{n} X_{i,h}X_{k,h} - \bar{X}_{i,\cdot}\bar{X}_{k,\cdot}.$$

Therefore

$$Z_n(i,k) = \frac{\sum_{h=1}^{n} X_{i,h}X_{k,h} - n\sigma_{i,k}}{\sqrt{n}} - (\sqrt{n}\bar{X}_{i,\cdot})\bar{X}_{k,\cdot}.$$

Here the first factor in the second term to the right is bounded because of the central limit theorem, and the second factor tends to zero in probability because of the law of large numbers. It also follows from the central limit theorem that the first term is asymptotically normally distributed with expectation 0 and covariances

$$\text{Cov}(X_{i,1}X_{k,1}, X_{j,1}X_{l,1}) = E(X_{i,1}X_{k,1}X_{j,1}X_{l,1}) - E(X_{i,1}X_{k,1})E(X_{j,1}X_{l,1}) =$$
$$E(X_{i,1}X_{k,1}X_{j,1}X_{l,1}) - \sigma_{i,k}\sigma_{j.l}.$$

It remains to show that

$$E(X_iX_kX_jX_l) = \sigma_{i,j}\sigma_{k,l} + \sigma_{i,l}\sigma_{k,j} + \sigma_{i,k}\sigma_{j,l}.$$

Let

$$\Gamma(\theta) = Ee^{\theta \cdot \boldsymbol{X}}$$

denote the moment generating function of $\boldsymbol{X} = (X_1, ..., X_m)$. Here $\theta = (\theta_1, ..., \theta_m)$. The right-hand side has the Taylor expansion

$$Ee^{\theta \cdot \boldsymbol{X}} = 1 + E(\theta \cdot \boldsymbol{X}) + \frac{1}{2}E(\theta \cdot \boldsymbol{X})^2 + \frac{1}{6}E(\theta \cdot \boldsymbol{X})^3 + \frac{1}{24}E(\theta \cdot \boldsymbol{X})^4 + \cdots$$

On the other hand, the left-hand side satisfies

$$\Gamma(\theta) = e^{\frac{1}{2}\theta \cdot Q\theta} = 1 + \frac{1}{2}\theta \cdot Q\theta + \frac{1}{2}(\frac{1}{2}\theta \cdot Q\theta)^2 + \ldots.$$

The fourth-order terms must coincide, and hence

$$\frac{1}{24}E(\theta \cdot \boldsymbol{X})^4 = \frac{1}{2}(\frac{1}{2}\theta \cdot Q\theta)^2;$$

that is,

$$E(\theta \cdot \boldsymbol{X})^4 = 3(\theta \cdot Q\theta)^2.$$

Differentiating both sides, we get

$$\frac{\partial^4}{\partial\theta_i\partial\theta_j\partial\theta_k\partial\theta_l}E(\theta \cdot \boldsymbol{X})^4 = 4 \cdot 3 \cdot 2 \cdot 1 E(X_iX_kX_jX_l),$$

and

$$\frac{\partial^4}{\partial\theta_i\partial\theta_j\partial\theta_k\partial\theta_l}(\theta \cdot Q\theta)^2 = 8(\sigma_{i,j}\sigma_{k,l} + \sigma_{i,l}\sigma_{j,k} + \sigma_{i,k}\sigma_{j,l}).$$

□

# CHAPTER 5

# TRADING STRATEGIES WITH CLOCK TIME HORIZON

The asymptotic result in Chapter 3 concerns trading strategies where the number of trades is given in advance. In this chapter the trading does not stop after a certain number of trades, but at a certain time. Here we need (for the first time) a stochastic model to relate this number to this time. This is the way we derive the Black–Scholes formula.

We also derive this formula in a more conventional way via the Black–Scholes equation. The reason for this is that this technique can be used to handle trading strategies for several assets (portfolios), and not just cash and a stock.

## 5.1 CLOCK TIME HORIZON

We shall steer the value of the portfolio toward $f(S_T)$ at time $T$. We thus need to determine $N$ such that $t_N \approx T$. (Or more exactly: $t_N \leq T < t_{N+1}$.) The level $\delta$ is chosen by us. Given this, $N$ is determined by the movements of the stock.

Let $N(\tau, t) = N_\delta(\tau, t)$ denote the number of $t_k$ in the interval $(\tau, t]$ and put $N(t) = N(0, t)$. The value of the portfolio at time $t$ and the stock price $S_t = s$ then equals

$$F_{N(t)}(s) = E[f(se^{Y_{N(t,T)}})|N(t,T)]. \tag{5.1}$$

This is a random variable that depends on the number $N(t, T)$ which is unknown at time $t$. Also the number of stocks

$$\frac{F_{N(t)}(se^{\delta}) - F_{N(t)}(se^{-\delta})}{se^{\delta} - se^{-\delta}} \tag{5.2}$$

is random. We shall, however, show that the random fluctuations becomes negligible as $\delta \to 0$.

Define $L_t$ by $S_t = S_0 e^{L_t}$, – that is, $L_t = \ln(S_t/S_0)$ – and define the volatility $\sigma$ by $\sigma^2 = \text{Var}(L_1)$.

The function $F(t, s)$ is defined by

$$F(t,s) = e^{-r(T-t)} Ef(s\exp[(r - \frac{\sigma^2}{2})(T-t) + \sigma\sqrt{T-t}Z]), \tag{5.3}$$

where $Z$ is normally distributed $(0, 1)$.

The interest rate is assumed to be constant, $= r$. In the case when the interest rate equals 0, the next theorem follows from Theorem 3.5.

**Theorem 5.1.** *Assume that*

$$\delta^2 N_\delta(t, T) \to \sigma^2(T - t) \tag{5.4}$$

*in probability as $\delta \to 0$. Then the portfolio value (5.1) tends to $F(t, s)$, and the number of stocks (5.2) tends to $\frac{\partial F(t,s)}{\partial s}$ in probability as $\delta \to 0$.*

*Proof.* We know that the formula holds when $r = 0$ and therefore when we replace the values of the stock and the portfolio with the present values $\bar{S}_t = e^{-rt}S_t$ respectively

$$\bar{F}(t, \bar{S}_t) = e^{-rt}F(t, e^{rt}\bar{S}_t).$$

The general case follows from this observation. Note that

$$\bar{f}(\bar{s}_t x) = e^{-rT}f(e^{rT}\bar{s}_t x) = e^{-rT}f(e^{r(T-t)}s_t x).$$

□

Here we shall show that the convergence (5.4), and hence also the conclusion of Theorem 5.1, holds under the assumption that the stock price follows Black–Scholes model, (4.3).

**Theorem 5.2.** *If the stock price follows the Black–Scholes model, then the convergence (5.4) holds.*

We shall give a heuristic argument below if you think the following proof is too technical.

*Proof.* Let $\tau_k = t_k - t_{k-1}$ for $k = 1, 2, ....$ These random variables are independent with the common distribution

$$F_\delta(t) = \text{Prob}(\max_{0\le s\le t} |\nu s + \sigma W(s)| \ge \delta) = G_{\nu\delta/\sigma^2}(t\sigma^2/\delta^2),$$

where

$$G_\epsilon(t) = \text{Prob}(\max_{0\le u\le t} |\epsilon s + W(s)| \ge 1).$$

This is so because

$$\max_{0\le s\le t} |\nu s + \sigma W(s)|/\delta$$

has the same distribution as

$$\max_{0\le s\le t} |(\nu/\delta)s + W(\sigma^2 s/\delta^2)|.$$

(Let $u = \sigma^2 s/\delta^2$.)

Also $G_\epsilon(s) \to G_0(s)$ as $\epsilon \to 0$. It follows that

$$\text{Prob}(\sigma^2\tau_k/\delta^2 \le t) \to G_0(t)$$

as $\delta \to 0$.

We have

$$\tau_1 + \cdots + \tau_{N_\delta(T-t)} \le T - t \le \tau_1 + \cdots + \tau_{N_\delta(T-t)} + \tau_{N_\delta(T-t)+1}$$

that is,

$$\frac{\tau_1\sigma^2/\delta^2 + \cdots + \tau_{N_\delta(T-t)}\sigma^2/\delta^2}{N_\delta(T-t)} \le \frac{(T-t)\sigma^2/\delta^2}{N_\delta(T-t)} \le$$
$$\frac{\tau_1\sigma^2/\delta^2 + \cdots + \tau_{N_\delta(T-t)+1}\sigma^2/\delta^2}{N_\delta(T-t)}.$$

The extreme members tend to $\mu = \int tG_0(dt)$ as $\delta \to 0$ and hence

$$\delta^2 N_\delta(T-t) \to \sigma^2(T-t)/\mu.$$

An explicit expression for $G_0(t)$ is known and a consequence of this is $\mu = 1$. But this is overkill. A direct proof follows here.

Let $t_1, t_2, ..., \tau_1, \tau_2, ...$ and $N(t)$ be as above but with $\nu = 0$, $\sigma = 1$ and $\delta = 1$. Then

$$\tau_1 + \cdots + \tau_{N(t)} \le t < \tau_1 + \cdots + \tau_{N(t)+1}.$$

Taking expectations, we get

$$EN(t)E\tau \leq t < (EN(t)+1)E\tau, \quad \text{i.e.,} \quad \frac{t}{EN(t)+1} < E\tau \leq \frac{t}{EN(t)}.$$

Define $X_k = W_{t_k} - W_{t_{k-1}}$ for $k = 1, 2, ..., N(t)$, and $Y = W_t - W_{t_{N(t)}}$. Then $Y, X_1, X_2, ...$ are independent, and $X_k$ take the values $\pm 1$ with probability $\frac{1}{2}$, and $|Y| < 1$. We have

$$W_t = \sum_{k=1}^{N(t)} X_k + Y$$

and hence

$$t = EW_t^2 = EN(t)EX^2 + EY^2 = EN(t) + EY^2.$$

Therefore $t - 1 < EN(t) \leq t$. Combining this with the earlier inequality, we get

$$\frac{t}{t+1} < E\tau < \frac{t}{t-1}, \quad \text{for all } t > 1.$$

Therefore $E\tau = 1$. □

Here is a heuristic argument: We have $|\nu\tau_1 + \sigma W(\tau_1)| = \delta$. Assume that $\nu = 0$. Then $W(\tau_1)^2 = \delta^2/\sigma^2$, and hence

$$\delta^2/\sigma^2 = EW(\tau_1)^2 = EE(W(\tau_1)^2|\tau_1) =$$

$$EE(W(1)^2\tau_1|\tau_1) = EW(1)^2E\tau_1 = E\tau_1.$$

Also

$$T \approx \tau_1 + \cdots + \tau_{N_\delta(T)} \approx N_\delta(T)E\tau_1.$$

Therefore

$$T \approx N_\delta(T)\delta^2/\sigma^2.$$

## 5.2 BLACK–SCHOLES PRICING FORMULAS

We shall here derive formulas for the price at time zero of European call and put options on a nondividend-paying stock.

Let $T$ stand for the time to maturity. We shall write

$$d_\pm = \frac{l}{\bar{\sigma}} \pm \frac{\bar{\sigma}}{2}, \tag{5.5}$$

where

$$l = \ln(S_0/\bar{K}), \quad \bar{K} = e^{-rT}K \quad \text{and} \quad \bar{\sigma} = \sigma\sqrt{T}. \tag{5.6}$$

Thus $\bar{K}$ stands for the present value of the strike price, and $\bar{\sigma}$ for the volatility under the remaining time to maturity. We shall write

$$\phi(x) = \frac{1}{\sqrt{2\pi}}e^{-\frac{x^2}{2}} \quad \text{and} \quad \Phi(x) = \int_{-\infty}^{x} \phi(y)\,dy$$

for the density and distribution function of a standardized normal random variable.

### *Black–Scholes Call Option Formula*

$$C_0 = S_0\Phi(d_+) - \bar{K}\Phi(d_-). \tag{5.7}$$

It follows from Theorem 5.1 and (5.18) that the stock holding and the cash in the trading strategy that hedges a short position in a call are $S_0\Phi(d_+)$ and $-\bar{K}\Phi(d_-)$, respectively.

The stock price, the strike price, the expiration date, and the interest rate are known quantities. The volatility can be estimated using historical data. Formula (5.7) can therefore be used to calculate the price $C_0$.

Conversely, if all these quantities except the volatility is known, then the volatility can be solved from identity (5.7) with $C_0$ the actual price. This solution is known as the *implicit volatility*, and it does not have to coincide with the historical volatility. On the contrary, the implicit volatility usually increases with the distance between the stock price and the strike price. This phenomenon is known as the *volatility smile*.

*Proof of the formula.* The price is given by (5.3) with $f(s) = \max(0, s - K)$, and hence

$$C_0 = E[\max(0, S_0e^{-\frac{\bar{\sigma}^2}{2}+\bar{\sigma}Z} - \bar{K})] = e^{-\frac{\bar{\sigma}^2}{2}}S_0E[\max(0, e^{\bar{\sigma}Z} - A)], \tag{5.8}$$

where

$$A = \bar{K}e^{\frac{\bar{\sigma}^2}{2}}/S_0.$$

It follows from Exercise 5.1 below that the last expectation in (5.8) equals

$$e^{\frac{\bar{\sigma}^2}{2}}\Phi(\bar{\sigma} - D) - A\Phi(-D).$$

Here we used the identity $1 - \Phi(x) = \Phi(-x)$. The pricing formula (5.7) now follows from the identities

$$D = \frac{\ln A}{\bar{\sigma}} = -d_-, \quad \bar{\sigma} - D = d_+.$$

□

EXERCISE 5.1. Show that

$$\int_D^\infty e^{\sigma z}\phi(z)dz = e^{\sigma^2/2}(1-\Phi(D-\sigma))$$

and hence that

$$E[\max(0, e^{\sigma Z} - A)] = e^{\sigma^2/2}(1-\Phi(D-\sigma)) - A(1-\Phi(D)), \tag{5.9}$$

where $D = \frac{\ln A}{\sigma}$.

The two extreme cases are $\bar{\sigma} = 0$ and $\bar{\sigma} = \infty$.

If $\bar{\sigma}$ is close to 0, then $d_\pm \approx \infty$ if $S_0 > \bar{K}$, and $d_\pm \approx -\infty$ if $S_0 < \bar{K}$, and hence $C_0 \approx \max(0, S_0 - \bar{K})$. So in this case

$$C_T \approx e^{rT} C_0.$$

A short position can thus be hedged simply by keeping the premium $C_0$ in cash.

If, on the other hand, $\bar{\sigma}$ is large, $\bar{\sigma} \approx \infty$, then $d_\pm \approx \pm\infty$, and hence $C_0 \approx S_0$. A short position can be hedged simply by using the premium to buy the stock at once.

Another special case is when $S_0 = \bar{K}$. Then $l = 0$ and hence

$$C_0 = S_0(\Phi(\frac{\bar{\sigma}}{2}) - \Phi(-\frac{\bar{\sigma}}{2})) = S_0 \int_{-\frac{\bar{\sigma}}{2}}^{\frac{\bar{\sigma}}{2}} \phi(x)\,dx.$$

We have

$$\phi(0) > \phi(x) > \phi(0)e^{-\frac{\bar{\sigma}^2}{8}}$$

for $-\frac{\bar{\sigma}}{2} < x < \frac{\bar{\sigma}}{2}$, and hence

$$\frac{S_0\bar{\sigma}}{\sqrt{2\pi}} > C_0 > \frac{S_0\bar{\sigma}}{\sqrt{2\pi}} e^{-\frac{\bar{\sigma}^2}{8}}.$$

If, for example, $\sigma = 0.5$ and $T = 1/6$ (two months) then

$$e^{-\frac{\bar{\sigma}^2}{8}} = 0.995.$$

Also $1/\sqrt{2\pi} = 0.399$. Therefore we have the approximation

$$C_0 \approx S_0\bar{\sigma}0.40 \tag{5.10}$$

when $S_0 \approx \bar{K}$.

We next turn to the pricing formula for a put.

### *Black–Scholes Put Option Formula*

$$P_0 = -S_0\Phi(-d_+) + \bar{K}\Phi(-d_-). \tag{5.11}$$

In this case the stock holding and the cash in the trading strategy that hedges a short position are $-S_0\Phi(-d_+)$ and $\bar{K}\Phi(-d_-)$, respectively. Thus in contrast to a call, we have a long position in cash and a short in the stock.

EXERCISE 5.2. Show that (5.11) follows from (5.7) and the put-call parity.

## 5.2.1 Sensitivity to Perturbations

The expiration date(s) and the strike price(s) of an options portfolio are fixed, but the stock price, the interest rate, and the volatility may vary with time. Here we shall consider the change of the portfolio value during a short time interval $(t, t + \partial t)$. To begin with, we shall assume that the interest rate and the volatility are constant.

Let $F_t$ denote the value of the portfolio at time $t$. Recall (Exercise 4.16) that $\partial S_t = S_{t+\partial t} - S_t$ is of the order $\sqrt{\partial t}$ and that $(\partial S_t)^2$ is of the same order as $\partial t$. Neglecting terms of smaller order than that, Taylor's formula gives

$$\partial F = \frac{\partial F}{\partial t}\partial t + \frac{\partial F}{\partial s}\partial S_t + \frac{1}{2}\frac{\partial^2 F}{\partial s^2}(\partial S_t)^2. \tag{5.12}$$

The effect on the portfolio value of a change of the stock price is measured by the *delta* of the portfolio

$$\Delta = \frac{\partial F}{\partial s}, \tag{5.13}$$

and to some degree also the *gamma*

$$\Gamma = \frac{\partial^2 F}{\partial s^2}. \tag{5.14}$$

This term is the second-order approximation of the influence of a change of the stock price. The gamma is also the derivative of $\Delta$ with respect to $s$. It shows how fast the delta will change with the stock price.

The *theta*

$$\Theta = \frac{\partial F}{\partial t} \tag{5.15}$$

is the rate of change of the portfolio value with respect to time, everything else being equal.

The formula (5.12) can thus also be written

$$\partial F = \Delta\partial S_t + \frac{1}{2}\Gamma(\partial S_t)^2 + \Theta\partial t. \tag{5.16}$$

The price of a call at time $t$, $C_t$, is given by (5.7) with $S_0$ replaced by $s = S_t$ and $T$ replaced by $T - t$. The next exercise is useful when calculating the partial derivatives of $C_t$.

EXERCISE 5.3. Show that the following identity holds for all positive numbers $s$, $k$, and $\sigma$:

$$s\phi(\frac{\ln(s/k)}{\sigma} + \frac{\sigma}{2}) = k\phi(\frac{\ln(s/k)}{\sigma} - \frac{\sigma}{2}). \tag{5.17}$$

We shall now calculate the delta of a call:

$$\frac{\partial C}{\partial s} = \Phi(d_+) + s\phi(d_+)\frac{\partial d_+}{\partial s} - \bar{K}\phi(d_-)\frac{\partial d_-}{\partial s}.$$

We have

$$\frac{\partial d_+}{\partial s} = \frac{\partial d_-}{\partial s},$$

and by Exercise 5.3; $s\phi(d_+) = \bar{K}\phi(d_-)$. Therefore

$$\Delta = \Phi(d_+). \tag{5.18}$$

EXERCISE 5.4. Show that the following holds for a call:

(a) $\Gamma = \frac{\phi(d_+)}{s\sigma\sqrt{T-t}}$.

(b) $\Theta = -e^{-r(T-t)}K(r\Phi(d_-) + \frac{\sigma\phi(d_-)}{2\sqrt{T-t}})$

The rate of change with respect to the volatility is called the *vega* of the portfolio, and the rate of change with respect to the interest rate is called the *rho*:

$$\vartheta = \frac{\partial F}{\partial \sigma}, \quad \rho = \frac{\partial F}{\partial r}.$$

EXERCISE 5.5. Show that the following holds for a call:

(a) $\vartheta = e^{-r(T-t)}K\phi(d_-)\sqrt{T-t}$.

(b) $\rho = e^{-r(T-t)}K\Phi(d_-)(T-t)$.

The value of a call is thus an increasing function of both the volatility and the interest rate.

The sizes of $\Gamma$ and $\Theta$ can thus be considerable when the remaining time to maturity is short and $S_t \approx K$. On the other hand, both $\rho$ and $\vartheta$ will be small when the time to maturity is short.

Note that the value of all these parameters except for $\Delta$ will depend on whether we give the prices in Euro or cents. The parameters $\Delta$, $S\Gamma$, $\Theta/S$, $\vartheta/S$, and $\rho/S$ are free from this dependence of the choice of unit:

$$\frac{\partial C}{S} = \Delta \frac{\partial S}{S} + \frac{1}{2} S\Gamma (\frac{\partial S}{S})^2 + (\Theta/S)\partial t + (\vartheta/S)\partial\sigma + (\rho/S)\partial r.$$

EXERCISE 5.6. Assume that $S_t = \bar{K} = 10$ EUR, $\sigma = 0.40$, and $r = 0.04$. Calculate $\Delta$, $S\Gamma$, $\Theta/S$, $\vartheta/S$, and $\rho/S$ for a call in the three cases $T - t = 1, 1/4$, and $1/50$.

In the case of a put, the above quantities can be obtained from the corresponding quantities for a call and from the put-call parity.

EXERCISE 5.7. Show that the following holds for a put:

(a) $\Delta = -\Phi(-d_+)$.

(b) $\Gamma = \frac{\phi(d_+)}{s\sigma\sqrt{T-t}}$.

(c) $\Theta = e^{-r(T-t)} K (r\Phi(-d_-) - \frac{\sigma\phi(d_-)}{2\sqrt{T-t}})$

(d) $\vartheta = e^{-r(T-t)} K\phi(d_-)\sqrt{T-t}$.

(e) $\rho = -e^{-r(T-t)} K\Phi(-d_-)(T-t)$.

EXERCISE 5.8. A stock has the volatility $40\%$, and today the price is 12 EUR.

(a) Use the Black–Scholes formula to calculate the price of a call with strike price 10 when the remaining time to maturity is 3 months and the interest rate equals $4\%$.

(b) How will the price of the option change when the stock price changes with $\partial S$ during the day? Determine $A$ so that $\partial C/C = A\partial S/S$.

EXERCISE 5.9. Today the price of a put with strike price 10 and time to expiration 1/2 year is 0.830 EUR. The price of the stock is 9.75 and the interest rate equals $5\%$.

(a) What is the price of a call with the same strike price and expiration date?

(b) Estimate the implicit volatility.

EXERCISE 5.10. (Long gamma) Consider the portfolio that consists of a short position in $x$ shares of a stock and a long position in a call on that stock.

(a) Determine $x$ so that the delta of the portfolio equals 0.

(b) Assume that the volatility and the interest rate remains constant. What is the change of the portfolio value during the time interval $(0, \partial t)$, neglecting terms of smaller order than $\partial t$?

### 5.2.2 Hedging a Written Call

A short position in a call can be hedged by the trading strategy that has the target function $f(s) = \max(0, s - K)$. Here we shall use formulas (5.7) and (5.18) to do this. Here is a recipe.

***Recipe 5.3.***

Trade at $t_0 = 0, t_1, t_2, ..., t_{n-1}$. Let $C_k$ and $S_k$ stand for the prices of the call and the stock at time $t_k$, and let $\Delta_k$ be the delta of the call at that time.

At $t = 0$. Receive the premium $C_0$, calculate $\Delta_0$, borrow $A_0 = \Delta_0 S_0 - C_0$ EUR, and buy $\Delta_0$ stocks.

At $t = t_1$. The portfolio is now worth $\Delta_0 S_1 - A_0 e^{rt_1}$. Buy $\Delta_1 - \Delta_0$ stocks. The stock holding is now worth $\Delta_1 S_1$ and the cash is worth $-A_1$, where $A_1 = A_0 e^{rt_1} + (\Delta_1 - \Delta_0) S_1$.

Continue in this way.

At time $t_k$ the stock holding is $\Delta_{k-1} S_k$ and the cash is $-A_{k-1} e^{r(t_k - t_{k-1})}$. Buy then $\Delta_k - \Delta_{k-1}$ stocks. The amount cash is now $-A_k$, where

$$A_k = A_{k-1} e^{r(t_k - t_{k-1})} + (\Delta_k - \Delta_{k-1}) S_k.$$

Thus $P_k = \Delta_{k-1} S_k - A_{k-1} = \Delta_k S_k - A_k$ for $k = 1, 2, ..., n$, where $P_k$ denotes the portfolio value at time $t_k$.

The trading moments $t_1, t_2, ...$ can be as before, but also deterministic; $t_k = k\partial t$. In the next exercise you shall trade once a day.

EXERCISE 5.11. A call on a stock has strike price $25$ EUR, and the remaining time to maturity is one month. Today the price of the option is $0.82$, and the stock $25.1$. Make a portfolio that hedges a short position in the call in the case where the closing prices for the stock becomes $25.6, 25.8$, and $25.6$ during the following three days. The interest rate is $3.8\%$, and the volatility is $28\%$ per year.

### 5.2.3 Three Options Strategies Again

Consider again the three options strategies of Section 2.3.3. One can use formulas (5.7) and (5.11) for calls and puts to divide the portfolio values (2.9), (2.10), and (2.11) in cash and stock holding. The result is given in the following table:

| Cash | Stock holding |
|---|---|
| $\bar{K}\Phi(-d_-(K))/(1+p)$ | $S\Phi(d_+(K))/(1+p)$ |
| $\bar{K}\Phi(-d_-(K)) + \bar{H}\Phi(d_-(H))$ | $S\big(\Phi(d_+(K)) - \Phi(d_+(H))\big)$ |
| $\bar{H}\Phi(d_-(H))/(1-c)$ | $S\Phi(-d_+(H))/(1-c)$ |

Here we have marked the strike prices to be able to distinguish between $d_\pm(K)$ and $d_\pm(H)$.

EXERCISE 5.12. Show this.

A common feature of these strategies is that both the cash and the number of stocks are positive.

The quantity $d_+$ is an increasing function of $s$. It is therefore seen from the expressions for the number of stocks that when replicating the first strategy, one has to buy stocks when the stock price rises but should sell when it decreases. For the third strategy the opposite holds.

In the case of the second strategy we have

$$\frac{\partial}{\partial s}\big(\Phi(d_+(K)) - \Phi(d_+(H))\big) = \frac{\phi(d_+(K)) - \phi(d_+(H))}{s\bar{\sigma}}.$$

This derivative is positive if $d_+(K)^2 < d_+(H)^2$.

EXERCISE 5.13. Show that

$$d_+(K) = d_+(\sqrt{KH}) + \frac{\ln(H/K)}{2\bar{\sigma}},\ d_+(H) = d_+(\sqrt{KH}) - \frac{\ln(H/K)}{2\bar{\sigma}}$$

and therefore

$$d_+(K)^2 - d_+(H)^2 = 2d_+(\sqrt{KH})\frac{\ln(H/K)}{\bar{\sigma}}.$$

It follows that the number of stocks for the second portfolio is an increasing function of the stock price if $d_+(\sqrt{KH}) < 0$, that is,

$$S_t < \sqrt{KH}e^{-(r+\sigma^2/2)(T-t)}. \tag{5.19}$$

When the stock price is below the value to the right, it is as for the first strategy: Buy when the price increases and sell when it decreases. When the price is higher than the value to the right; it is as for the third strategy: Do the contrary. The number of stocks is thus maximal when we have equality in (5.19).

EXERCISE 5.14. Determine the value of this maximum.

In the next example we use the formulas (5.7) and (5.11) for the prices of calls and puts to illustrate the three strategies numerically.

***Example 5.4.*** Assume that $T = 1/4$ (three months), $r = 0.04$, and $\sigma = 0.40$. Let $K = 0.950S_0$, then $H = 1.087S_0$ for the second strategy. We shall use the same $H$ also for the third strategy. In this case $c = p = 0.051$.

The first strategy may be safe, but it is expensive. The price of the asset must raise $p = 5\%$ to get the money back, and at worst the loss will be $1 - (K/S_0)/(1+p) = 10\%$.

The second strategy will thus at worst give a loss of $1 - K/S_0 = 5\%$ and at most a gain of $H/S_0 - 1 = 9\%$.

The third strategy will at most yield $(H/S_0)/(1-c) - 1 = 14\%$, and it will give the money back even if the asset falls $5\%$.

## 5.3 THE BLACK–SCHOLES EQUATION

In this section we shall assume that the stock price, $S_t$, develops according to the Black–Scholes model; (4.3), and derive a differential equation for the portfolio value of the trading strategy that has the target value $f(S_T)$ at time $T$. We already know that the portfolio value is given by (5.3) and, hence, that this is the solution of the differential equation. Thus, we will not obtain any new result, but the method of this section will be used in the next section in connection with trading strategies for portfolios that consist of several assets.

As before, we shall consider a portfolio that consists of cash and a number of a stock. No money is brought into or taken away from the portfolio; the portfolio is *self-financing*.

The portfolio is rebalanced at

$$0 = t_0 < t_1 < \cdots < t_n = T.$$

Here $t_k = k\partial t$, where $\partial t = T/n$. The portfolio value at time $T$ shall be close to $f(S_T)$. Here $f(s)$ is a given target function. Here "close to" means that we can get the portfolio value arbitrarily close to the target if the lengths of the time intervals, $\partial t = t_k - t_{k-1}$, are made sufficiently small.

By considering the present values of the portfolio and the stock, we can and shall assume that the interest rate equals 0.

Let $P_t$ denote the value of the portfolio at time $t$, and let $c_{k-1}$ and $a_{k-1}$ be the amount of cash and the number of assets (stocks) in the time interval $[t_{k-1}, t_k)$, respectively. Thus

$$P_t = c_{k-1} + a_{k-1}S_t \quad \text{for } t_{k-1} \le t < t_k, \quad k = 1, 2, ..., n.$$

This identity holds also for $t = t_k$ since the portfolio is self-financing, and therefore

$$\partial P_{t_k} = a_{k-1}\partial S_{t_k}.$$

Here $\partial P_{t_k} = P_{t_k} - P_{t_{k-1}}$ and $\partial S_{t_k} = S_{t_k} - S_{t_{k-1}}$.

The development of the stock, the portfolio value at time 0, and the number of stocks $a_0, a_1, ...$ therefore determine the values of the portfolio by the relation

$$P_{t_k} = P_0 + a_0\partial S_{t_1} + a_1\partial S_{t_2} + \cdots + a_{k-1}\partial S_{t_k}.$$

We shall determine $P_0$ and $a_0, a_1, \ldots$ such that $P_T$ is close to $f(S_T)$. In order to do this we shall try to find a function, $F(t,s)$, $0 \le t \le T$, $s > 0$, which is such that $F(t, S_t)$ can be approximated by $P_t$.

We have

$$F(t_k, S_{t_k}) = F(0, S_0) + \partial F_{t_1} + \partial F_{t_2} + \cdots + \partial F_{t_k},$$

where

$$\partial F_{t_k} = F(t_k, S_{t_k}) - F(t_{k-1}, S_{t_{k-1}})$$

and therefore

$$F(t_k, S_{t_k}) - P_{t_k} = F(0, S_0) - P_0 + D_1 + D_2 + \cdots + D_k$$

where

$$D_k = \partial F_{t_k} - a_{k-1}\partial S_{t_k}.$$

We shall assume that $f$ is a regular function.

EXERCISE 5.15. Use a Taylor expansion of $F$, along with Exercise 4.16, to show that

$$D_k = \Big(\frac{\partial}{\partial s}F(t_{k-1}, S_{t_{k-1}}) - a_{k-1}\Big)\partial S_{t_k} +$$

$$\frac{\partial}{\partial t}F(t_{k-1}, S_{t_{k-1}})\partial t + \frac{1}{2}\frac{\partial^2}{\partial s^2}F(t_{k-1}, S_{t_{k-1}})(\partial S_{t_k})^2 + O(\sqrt{\partial t}^3).$$

Note that

$$\partial S_{t_k} = S_{t_{k-1}}(e^{\nu\partial t + \sigma\partial W_{t_k}} - 1)$$

and that $\partial W_{t_k} = Z_k\sqrt{\partial t}$, where $Z_k$ is normal (0,1). Therefore

$$\partial S_{t_k} = O(\sqrt{\partial t})$$

and

$$(\partial S_{t_k})^2 = S_{t_{k-1}}^2\sigma^2 Z_k^2 \partial t + O(\sqrt{\partial t}^3).$$

The random variable $Z_k^2$ has expectation 1. Therefore put

$$\epsilon_k = e(t_{k-1}, S_{t_{k-1}})(Z_k^2 - 1)\partial t,$$

where

$$e(t,s) = \frac{s^2\sigma^2}{2}\frac{\partial^2}{\partial s^2}F(t,s)$$

We thus have

$$D_k = (\frac{\partial}{\partial s}F(t_{k-1}, S_{t_{k-1}}) - a_{k-1})\partial S_{t_k} +$$

$$(\frac{\partial}{\partial t}F(t_{k-1}, S_{t_{k-1}}) + \frac{1}{2}S_{t_{k-1}}^2\sigma^2\frac{\partial^2}{\partial s^2}F(t_{k-1}, S_{t_{k-1}}))\partial t + \epsilon_k + O(\sqrt{\partial t}^3).$$

It follows that if $F(t, s)$ is a solution of the *Black–Scholes equation*

$$\frac{\partial F}{\partial t} + \frac{s^2\sigma^2}{2}\frac{\partial^2 F}{\partial s^2} = 0,$$

and if we choose

$$P_0 = F(0, S_0) \quad \text{and} \quad a_k = \frac{\partial}{\partial s}F(t_k, S_{t_k}) \quad \text{for } k = 0, 1, ..., n-1,$$

then

$$F(t_k, S_{t_k}) - P_{t_k} = \epsilon_1 + \epsilon_2 + \cdots + \epsilon_k + O(\sqrt{\partial t}).$$

We shall calculate the expectation and variance of the sum to the right. Let $E_t X$ denote the conditional expectation of $X$, given the stock prices up to time $t$.

We have

$$E\epsilon_k = EE_{t_{k-1}}\epsilon_k = \partial t Ee(t_{k-1}, S_{t_{k-1}})E_{t_{k-1}}(Z_k^2 - 1) = 0.$$

EXERCISE 5.16. Show that

(a) $E\epsilon_k\epsilon_j = 0$ for $k \neq j$.

(b) $E\epsilon_k^2 = (\partial t)^2 G(t_{k-1})E(Z_k^2 - 1)^2$, where $G(t) = Ee(t, S_t)^2$.

(c) $E(Z_k^2 - 1)^2 = 2$.

It follows that

$$\text{Var}(\epsilon_1 + \cdots + \epsilon_n) = 2\partial t \sum_{k=1}^{n} G(t_{k-1})\partial t =$$

$$2\partial t \int_0^T G(t)\, dt + O((\partial t)^2) = O(\partial t).$$

Here we approximated the Riemann sum by an integral. The next exercise gives a bound for the error.

EXERCISE 5.17. Show that

$$|\int_0^T G(t)\, dt - \sum_{k=1}^{n} G(t_{k-1})\partial t| \leq \partial t \int_0^T |G'(s)|\, ds.$$

**Theorem 5.5.** *Assume that $F(t, s)$ solves the Black–Scholes equation and that $F(T, s) = f(s)$ for all $s > 0$. Choose $P_0 = F(0, S_0)$ and $a_k = \frac{\partial F(t_k, S_{t_k})}{\partial s}$ for $k = 0, 1, ..., n-1$. Then*

$$E(P_T - F(S_T)) = O(\sqrt{\partial t}), \;\; E(P_T - F(S_T))^2 = O(\partial t)$$

*and hence $P_T - F(S_T) = O(\sqrt{\partial t})$ in probability.*

The above proof contains some technical details. If we peel these off and neglect terms of the order of magnitude $\sqrt{\partial t}^3$, then the proof looks like this:

1. A Taylor expansion and $\partial S = O(\sqrt{\partial t})$ give

$$\partial F = \frac{\partial F}{\partial t}\partial t + \frac{\partial F}{\partial s}\partial S + \frac{1}{2}\frac{\partial^2 F}{\partial s^2}(\partial S)^2.$$

2. Replace $(\partial W)^2$ in the expression $(\partial S)^2 = \sigma^2 S^2 (\partial W)^2$ by $\partial t$:

$$\partial F = \frac{\partial F}{\partial s}\partial S + (\frac{\partial F}{\partial t} + \frac{1}{2}\frac{\partial^2 F}{\partial s^2}\sigma^2 s^2)\partial t.$$

3. If $F$ solves the Black–Scholes equation and $a = \frac{\partial F}{\partial s}$, then $\partial F = a\partial S$ and therefore $F = \int dF = \int a\, dS = P$.

EXERCISE 5.18. Show that if the interest rate is constant $= r$, then the Black–Scholes equation takes the form

$$\frac{\partial F}{\partial t} + rs\frac{\partial F}{\partial s} + \frac{s^2\sigma^2}{2}\frac{\partial^2 F}{\partial s^2} = rF. \tag{5.20}$$

*Hint:* See the proof of Theorem 5.1.

***The Solution of the Equation*** We shall first show that one solution is given by (5.3). Therefore let

$$S_t = S_0 e^{-\frac{\sigma^2}{2}t + \sigma W_t},$$

where $W_t$ is a Wiener process and put

$$F(t, S_t) = E_t f(S_T) = E(f(S_T)|S_t).$$

Let $E_t$ stand for the conditional expectation given $S_t$. We have

$$E_t F(t+\partial t, S_{t+\partial t}) = E_t E(f(S_T)|S_{t+\partial t}) = E(f(S_T)|S_t) = F(t, S_t)$$

and therefore

$$0 = E_t(F(t+\partial t, S_{t+\partial t}) - F(t, S_t)) =$$

$$\frac{\partial F(t,S_t)}{\partial t}\partial t + \frac{\partial F(t,S_t)}{\partial s}E_t\partial S_t + \frac{1}{2}\frac{\partial^2 F(t,S_t)}{\partial s^2}E_t(\partial S_t)^2 + \cdots.$$

Applying Exercise 4.16 with $\nu = -\frac{\sigma^2}{2}$, we get

$$E_t\partial S_t = O((\partial t)^2),\ \ E_t(\partial S_t)^2 = S_t^2\sigma^2\partial t + O((\partial t)^2,\ \ \text{and}\ \ E_t(\partial S_t)^k = O((\partial t)^2)$$

for $k = 3, 4, \ldots$ .

It follows that $F(t, s)$ solves the Black–Scholes equation. It is also clear that

$$F(T, S_T) = f(S_T).$$

The uniqueness (that there is only one solution) follows from the theory for partial differential equations. We do not presuppose that this is known, but shall instead use an arbitrage argument: Assume that there are two solutions $F_1$ and $F_2$ and that $F_1(t_0, S_{t_0}) < F_2(t_0, S_{t_0})$. Replicate the portfolio $F_1(t_0, S_{t_0}) - F_2(t_0, S_{t_0})$. This gives the income $F_2(t_0, S_{t_0}) - F_1(t_0, S_{t_0}) > 0$ at $t = t_0$ and the value of the portfolio at $t = T$ equals 0. We have thus made arbitrage. The assumption that the market is free of arbitrage therefore implies uniqueness.

## 5.4 TRADING STRATEGIES FOR SEVERAL ASSETS

So far we have considered portfolios consisting of two assets; cash and a stock. Here we shall consider the more general situation when the portfolio consists of $m \geq 2$ assets. One of these can but does not have to be cash. We shall first proceed as in the previous section and treat the cash separately, and then give a symmetrical formulation where we consider cash as one among the other assets. In both cases we shall assume that Model B, (4.2), holds.

### 5.4.1 An Unsymmetrical Formulation

We thus have cash and $m - 1$ other assets. The (continuous) interest rate is assumed to be constant $= r$. As earlier, we shall consider the present values of the assets and therefore put $r = 0$.

The portfolio values satisfies

$$P_n(t) = c_{k-1} + a_1(k-1)S_1(t) + \cdots + a_{m-1}(k-1)S_{m-1}(t) \tag{5.21}$$

for $t_{k-1} \leq t < t_k$, $k = 1, ..., n$. Here $0 = t_0 < t_1 < \cdots < t_n = T$ and $\boldsymbol{S}(t) = (S_1(t), ..., S_{m-1}(t))$ are the prices of the assets at time $t$.

We shall steer the portfolio value toward $f(\boldsymbol{S}_T)$ at time $T$, where $f$ is a given *target function*: A continuous function $(0, \infty)^{m-1} \ni \boldsymbol{s} \to f(\boldsymbol{s}) \in \mathbb{R}$.

We can now proceed as earlier, but with the difference that we have a function of several variables:

$$F(T, \boldsymbol{S}_T) - P_T = \epsilon_1 + \epsilon_2 + \cdots + \epsilon_n + O(\sqrt{\partial t}).$$

This difference tends to zero if

$$P_n(0) = F(0, \boldsymbol{S}(0)) \quad \text{and} \quad a_i(k) = \frac{\partial}{\partial s_i} F(t_k, \boldsymbol{S}(t_k))$$

and if $F(t, \boldsymbol{s})$ solves the equation

$$\frac{\partial F}{\partial t} + \frac{1}{2} \sum_{i=1}^{m-1} \sum_{j=1}^{m-1} s_i s_j \sigma_{ij} \frac{\partial^2 F}{\partial s_i \partial s_j} = 0. \tag{5.22}$$

In this case

$$\epsilon_k = \sum_{i=1}^{m-1} \sum_{j=1}^{m-1} \frac{s_i s_j \sigma_{ij}}{2} \frac{\partial^2 f}{\partial s_i \partial s_j} (Z_i(k) Z_j(k) - \sigma_{ij}) \partial t,$$

where $Z_i(k) = (X_i(t_k) - X_i(t_{k-1}))/\sqrt{\partial t}$.

As earlier, $\epsilon_1,\dots,\epsilon_n$ are uncorrelated and have zero expectations.

The function $F(t, \boldsymbol{s})$ defined by

$$F(t, \boldsymbol{s})) = Ef(s_1 e^{-\frac{\sigma_{11}}{2}\tau + X_1(\tau)}, \dots, s_{m-1} e^{-\frac{\sigma_{m-1\,m-1}}{2}\tau + X_{m-1}(\tau)}),$$

where $\tau = T - t$ solves Eq. (5.22) and

$$F(T, \boldsymbol{s}) = f(\boldsymbol{s}). \tag{5.23}$$

The above formula thus holds if the interest rate equals zero.

The identity (5.3) now takes the form

$$F(t, \boldsymbol{s}) =$$

$$e^{-r\tau} Ef(s_1 e^{r\tau - \frac{\sigma_{11}}{2}\tau + X_1(\tau)}, \dots, s_{m-1} e^{r\tau - \frac{\sigma_{m-1\,m-1}}{2}\tau + X_{m-1}(\tau)}) \tag{5.24}$$

Even in this case we obviously have (5.23).

In this way we get the multidimensional version of Theorem 5.5.

**Theorem 5.6.** *Assume that $F$ is as in (5.24) and that $P_n(t)$ is as in (5.21) with*

$$a_i(k-1) = \frac{\partial}{\partial s_i} F(t_{k-1}, \boldsymbol{S}(t_{k-1})).$$

*If $P_n(0) = F(0, \boldsymbol{S}(0))$, then $P_n(t) - F(t, \boldsymbol{S}(t)) \to 0$ in probability, as $n \to \infty$ and* $\max_{1 \le k \le n}(t_k - t_{k-1}) \to 0$.

EXERCISE 5.19. Show that Eq. (5.22) takes the form

$$\frac{\partial F}{\partial t} + r \sum_{i=1}^{m-1} s_i \frac{\partial F}{\partial s_i} + \frac{1}{2} \sum_{i=1}^{m-1} \sum_{j=1}^{m-1} s_i s_j \sigma_{ij} \frac{\partial^2 F}{\partial s_i \partial s_j} = rF. \tag{5.25}$$

when the interest rate is constant $= r$.

### 5.4.2 A Symmetrical Formulation

Here we shall consider a portfolio that consists of $m$ assets, and we shall write $g$ and $G$ for the target function and the portfolio value. In this case, $g$ is a function of $m$ variables $\boldsymbol{s} = (s_1, ..., s_m)$, and $G$ has to satisfy

$$\frac{\partial G}{\partial t} + \frac{1}{2}\sum_{i=1}^{m}\sum_{j=1}^{m} s_i s_j \sigma_{ij} \frac{\partial^2 G}{\partial s_i \partial s_j} = 0 \tag{5.26}$$

and

$$G(T, \boldsymbol{s}) = g(\boldsymbol{s}). \tag{5.27}$$

This boundary value problem has the solution

$$G(t, \boldsymbol{s}) = Eg(s_1 e^{-\frac{\sigma_{11}}{2}\tau + X_1(\tau)}, ..., s_m e^{-\frac{\sigma_{mm}}{2}\tau + X_m(\tau)}). \tag{5.28}$$

The weights are given by

$$\boldsymbol{b}(t, \boldsymbol{s}) = \text{grad}_{\boldsymbol{s}} G(t, \boldsymbol{s}),$$

and hence we must have

$$G(t, \boldsymbol{s}) = \boldsymbol{b}(t, \boldsymbol{s}) \cdot \boldsymbol{s}. \tag{5.29}$$

We shall give a condition on $g$ that implies that (5.29) is satisfied.

***Definition 5.7.*** A *homogeneous target function* is a continuous function $(0, \infty)^m \ni \boldsymbol{s} \to g(\boldsymbol{s}) \in \mathbb{R}$ that satisfies

$$g(k\boldsymbol{s}) = kg(\boldsymbol{s}) \tag{5.30}$$

for all $k > 0$. Here $k\boldsymbol{s} = (ks_1, ..., ks_m)$.

It follows from (5.28) that the condition (5.30) implies that $G(t, k\boldsymbol{s}) = kG(t, \boldsymbol{s})$, and it follows from the lemma below that this in turn implies (5.29).

**Lemma 5.8.** *Let $H(\boldsymbol{s})$ be a differentiable function defined for $\boldsymbol{s} \in (0, \infty)^m$. Then*

$$H(\boldsymbol{s}) = \boldsymbol{s} \cdot \textit{grad}\, H(\boldsymbol{s}) \tag{5.31}$$

*for all $\boldsymbol{s} \in (0, \infty)^m$ if and only if*

$$H(k\boldsymbol{s}) = kH(\boldsymbol{s}) \tag{5.32}$$

*for all $\boldsymbol{s} \in (0, \infty)^m$, and all $k > 0$.*

*Proof.* Assume that (5.31) holds. Fix $\boldsymbol{s}$, and put $h(k) = H(k\boldsymbol{s})$. Then

$$h'(k) = \boldsymbol{s} \cdot \text{grad}\, H(k\boldsymbol{s}) = h(k)/k,$$

and hence $h(k) = h(1)k$.

Assume that (5.32) holds. Differentiate both sides with respect to $k$ and put $k = 1$.□

One starts at $t = t_0 = 0$ with the portfolio value $P_n(0) = \boldsymbol{b}(t_0, \boldsymbol{S}(t_0)) \cdot \boldsymbol{S}(t_0)$. At $t = t_1$ this portfolio is worth $P_n(t_1) = \boldsymbol{b}(t_0, \boldsymbol{S}(t_0)) \cdot \boldsymbol{S}(t_1)$. However, immediately after the rebalancing at $t = t_1$, it is worth $\boldsymbol{b}(t_1, \boldsymbol{S}(t_1)) \cdot \boldsymbol{S}(t_1)$, which in general is different from $P_n(t_1)$. This problem can be solved, for example, by adding a transaction account to the portfolio and proceed as before. The account will be negligible as one trades more and more often, that is, when $n \to \infty$.

The portfolio is constructed in the following way:

$$P_n(0) = \boldsymbol{b}(0, \boldsymbol{S}(0)) \cdot \boldsymbol{S}(0) \text{ and}$$

$$P_n(t) = P_n(t_{k-1}) + \boldsymbol{b}(t_{k-1}, \boldsymbol{S}(t_{k-1}) \cdot (\boldsymbol{S}(t) - \boldsymbol{S}(t_{k-1})) \tag{5.33}$$

for $t_{k-1} \leq t < t_k$, $k = 1, ..., n$. We therefore have the following theorem.

**Theorem 5.9.** *If $G$ is as in (5.28) and $P_n(t)$ is as in (5.33), then $P_n(t) - G(t, \boldsymbol{S}(t)) \to 0$ in probability, as $n \to \infty$, and* $\max_{1 \leq k \leq n}(t_k - t_{k-1}) \to 0$.

Assume that cash is asset no. $m$. Then the connection between the target functions $f$ and $g$ is

$$g(s_1, ..., s_m) = s_m f(\frac{s_1}{s_m}, ..., \frac{s_{m-1}}{s_m}),$$

$$f(s_1, ..., s_{m-1}) = g(s_1, ..., s_{m-1}, 1). \tag{5.34}$$

EXERCISE 5.20. Let $f$ be any continuous real-valued function defined in $(0, \infty)^{m-1}$, and define $g$ as in (5.34). Show that $g$ is a homogeneous target function.

We have $S_m(t) = S_m(0)e^{rt}$, and hence

$$G(t, \boldsymbol{S}(t)) = S_m(0)e^{rt} F(t, e^{-rt} \boldsymbol{S}'(t)/S_m(0)),$$

where $\boldsymbol{S}'(t) = (S_1(t), ..., S_{m-1}(t))$. Thus if we norm the last asset such that $S_m(0) = 1$ and put $t = T$, we get

$$e^{-rT} g(S(T)) = f(e^{-rT} S_1(T), ..., e^{-rT} S_{m-1}(T)).$$

The present value of the target can thus be chosen to be (almost) any continuous function of the present values of the asset prices.

The portfolio value can if necessary be computed from (5.28) using numerical integration. However, when implementing a trading strategy using (5.33), we have to compute the number of assets (5.29) rather than the portfolio value. The next lemma gives a formula for the asset holdings.

**Lemma 5.10.** *If $Q$ is strictly positive definite, then*

$$(s_1 \frac{\partial G(t, s)}{\partial s_1}, ..., s_m \frac{\partial G(t, s)}{\partial s_m}) =$$

$$E[g(s_1 e^{-\tau\frac{\sigma_{11}}{2}+X_1(\tau)}, ..., s_m e^{-\tau\frac{\sigma_{mm}}{2}+X_m(\tau)})(\tau Q)^{-1} X(\tau)]. \tag{5.35}$$

*Proof.* Let $x = (x_1, ..., x_m)$,

$$\phi(x) = \frac{1}{\sqrt{2\pi}^m \sqrt{\det Q}} e^{-\frac{x \cdot Q^{-1} x}{2}}, \quad \text{and} \quad \phi(\tau, x) = \phi(\frac{x}{\sqrt{\tau}}) \frac{1}{\sqrt{\tau}^m}.$$

Then

$$G(t, s) = \int g(s_1 e^{-\frac{\sigma_{11}}{2}\tau + \sqrt{\tau} x_1}, ..., s_m e^{-\frac{\sigma_{mm}}{2}\tau + \sqrt{\tau} x_m}) \phi(x)\, dx.$$

Make the substitution $y = l + \sqrt{\tau} x$, where $l_i = \ln s_i - \frac{\sigma_{ii}}{2}\tau$, and write $h(y) = g(e^{y_1}, ..., e^{y_m})$. This yields the representation

$$G(t, s) = \int h(y) \phi(\tau, y - l)\, dy.$$

We have $\frac{\partial (y_k - l_k)}{\partial s_i} = -\frac{\delta_{ki}}{s_i}$, where $\delta_{ii} = 1$ and $\delta_{ki} = 0$ for $k \neq i$. Therefore

$$\frac{\partial G}{\partial s_i} = \int h(y) \frac{\partial \phi(\tau, y - l)}{\partial y_i} (-\frac{1}{s_i})\, dy = \frac{1}{s_i} \int h(y) \phi(\tau, y - l)((\tau Q)^{-1}(y - l))_i\, dy,$$

and hence (5.35) holds. □

### 5.4.3 Examples

***Example 5.11.*** Call.

In the case $m = 2$ and $f(x) = \max(0, x - K)$, we get

$$g(S_1(T), S_2(T)) = \max(0, S_1(T) - e^{rT} S_2(0) K).$$

In this case, $g$ is thus the target function for a call with strike price $K$ if $S_2(0)$ is the present value at time 0 of one EUR at $T$, $S_2(0) = e^{-rT}$.

***Example 5.12.*** Buy and Hold.

This is the "trading strategy" for which the number of each asset is constant:

$$a_i(t, s) = P(0) v_i / S_i(0),$$

for all $t$ and $s$. Here $P(0)$ is the value of the portfolio at time 0, and $v_i$ is the weight of asset $i$ at time 0. Thus

$$G(t, s) = g(s) = P(0)(w_1 s_1 / S_1(0) + \cdots + w_m s_1 / S_m(0)),$$

and hence the portfolio value is proportional to a weighted arithmetic mean of $S_i(t)/S_i(0)$, $i = 1, ..., m$.

***Example 5.13.*** Fixed Weights.

In this case the portfolio is rebalanced in such a way that the weight of asset $i$ is a fixed proportion $v_i$, where $v_1 + \cdots + v_m = 1$. Thus

$$\frac{\partial G}{\partial s_i} s_i = v_i G,$$

and hence

$$G(t,s) = K(t) \prod_{i=1}^{m} (\frac{s_i}{S_i(0)})^{v_i}$$

and $g(s) = G(T, s)$.

It therefore follows from (5.28) that

$$K(t) \prod_{i=1}^{m} (\frac{s_i}{S_i(0)})^{v_i} = K(T) \prod_{i=1}^{m} (\frac{s_i}{S_i(0)})^{v_i} E[\prod_{i=1}^{m} (e^{-\tau \frac{\sigma_{ii}}{2} + X_i(\tau)})^{v_i}],$$

and hence

$$K(t) = K(T) e^{-\tau \frac{1}{2}(\sum_{i=1}^{m} \sigma_i^2 v_i)} E[e^{\boldsymbol{v} \cdot X(\tau)}] = K(T) e^{-\tau \frac{1}{2}(\sum_{i=1}^{m} \sigma_i^2 v_i)} e^{\tau \frac{\boldsymbol{v} \cdot Q \boldsymbol{v}}{2}}.$$

Therefore

$$K(t) = K(0) e^{Lt},$$

where $L = \frac{1}{2}(\frac{1}{2}(\sum_{i=1}^{m} \sigma_i^2 v_i) - \boldsymbol{v} \cdot Q\boldsymbol{v})$. Thus

$$G(t,s) = P(0) e^{Lt} \prod_{i=1}^{m} (\frac{s_i}{S_i(0)})^{v_i}. \tag{5.36}$$

The portfolio value is thus proportional to a weighted geometric mean of $S_i(t)/S_i(0)$, $i = 1, ..., m$.

***Example 5.14.*** Ordered Returns.

By considering $S_i(t)/S_i(0)$ instead of $S_i(t)$, we shall here assume that $S_i(0) = 1$ for $i = 1, ..., m$. Let $s_{(1)} \leq \cdots \leq s_{(m)}$ be $s_1, ..., s_m$ ordered in increasing order and let $g(s) =$ const. $s_{(r)}$, where the constant is chosen such that $G(0, \mathbf{1}) = P(0)$.

Thus $r = 1$ and $r = m$ gives the minimum and maximum, respectively. Another example is when $m$ is odd and $r = [m/2]$. In this case the target function is thc median.

Using the commando "sort" in Matlab, the number of assets can be computed by numerical integration of (5.35). Another possibility is to use Monte Carlo methods.

We shall now give explicit expressions for the portfolio value and the number of assets in the case $m = 2$. In this case we have the two target functions $g_1(s_1, s_2) = \max(s_1, s_2)$ and $g_2(s_1, s_2) = \min(s_1, s_2)$. Note that the first is the target function for the strategy Protective put, and the second is that for the Covered call in Section 2.3.3. In this case, one of the assets were cash, however.

Let $\phi_{Qt}(x_1, x_2)$ denote the density of $(X_1(t), X_2(t))$. We shall start with the first strategy. We have

$$G_1(t, s_1, s_2) = E \max(s_1 e^{-\frac{\sigma_{11}}{2}\tau + X_1(\tau)}, s_2 e^{-\frac{\sigma_{22}}{2}\tau + X_2(\tau)}) = s_1 I_{12} + s_2 I_{21},$$

where

$$I_{ij} = \int_{x_i - x_j > -\lambda_{ij}} e^{-\frac{\sigma_{ii}}{2}\tau + x_i} \phi_{Q\tau}(x_1, x_2)\, dx_1 dx_2,$$

and

$$\lambda_{ij} = \ln s_i - \ln s_j + \frac{\sigma_{ii} - \sigma_{jj}}{2}\tau.$$

It follows from Exercise 5.21 below that

$$I_{12} = \int_{y_2 - y_1 > \ln s_1 - \ln s_2 + \frac{\omega_{12}^2}{2}\tau} \phi_{Q\tau}(y_1, y_2)\, dx_1 dx_2,$$

where

$$\omega_{ij}^2 = \sigma_{ii} + \sigma_{jj} - 2\sigma_{ij}.$$

Note that if $(Y_1, Y_2)$ is normally distributed with zero expectation and covariance matrix $Q\tau$, then $Y_2 - Y_1$ is normally distributed with zero expectation and variance $\omega_{12}^2\tau$. Therefore

$$I_{ij} = \Phi(d_{ij}),$$

where

$$d_{ij} = \frac{\ln(s_i/s_j)}{\omega_{ij}\sqrt{\tau}} + \frac{\omega_{ij}\sqrt{\tau}}{2}.$$

It follows that

$$G_1(t, s_1, s_2) = s_1\Phi(d_{12}) + s_2\Phi(d_{21}).$$

In order to calculate the portfolio value for the other target function, $g_2$, we use the identities $\min(s_1, s_2) = s_1 + s_2 - \max(s_1, s_2)$, and

$$Ee^{-\frac{\sigma_{ii}}{2}\tau + X_i(\tau)} = 1 \text{ and } 1 - \Phi(x) = \Phi(-x).$$

These give

$$G_2(t, s_1, s_2) = s_1\Phi(-d_{12}) + s_2\Phi(-d_{21}).$$

EXERCISE 5.21. Show, for example by calculating the moment-generating function, that

$$e^{-\frac{\sigma_{11}}{2}\tau + x_1}\phi_{Q_\tau}(x_1, x_2) = \phi_{Q_\tau}(x_1 - \sigma_{11}\tau, x_2 - \sigma_{12}\tau).$$

EXERCISE 5.22. Calculate the number of the two assets in the two portfolios as a function of time and the asset prices.

*Hint:* Use Exercise 5.3.

The first strategy may look more tempting than the other, but the picture will be more balanced if we compare strategies that have the same value from the beginning.

EXERCISE 5.23. Assume that $S_1(0) = S_2(0) = 1$.

(a) Determine $c$ and $C$ such that $cG_1(0, 1, 1) = 1$ and $CG_2(0, 1, 1) = 1$. In particular: What are $c$ and $C$ if we have $T = 1$, $\sigma_{11} = 0.16$, $\sigma_{22} = 0.25$, and the correlation $\sigma_{12}/\sqrt{\sigma_{11}\sigma_{22}} = 0.35$?

(b) For which values of $S_1(T)$ and $S_2(T)$ will the strategies give a profit?

(c) For which values of $S_1(T)$ and $S_2(T)$ is the second strategy preferable to the first?

## 5.5 NOTES

Our derivation in Section 5.1 is not the usual one, but the same method has been used for numerical reasons by Rogers and Stapelton (1998).

A derivation of the Black–Scholes formula using martigales can be found in, for example, Baxter and Rennie (1996).

# CHAPTER 6

# DIVERSIFICATION

In this chapter we shall manage risk by forming portfolios. We shall start with the minimum variance portfolio, and then proceed with portfolios that have maximal growth under various constraints. It turns out that all optimal portfolios are linear combinations of three portfolios: cash, the minimum variance portfolio, and a certain auxiliary stock portfolio.

A trading strategy for several assets as considered in Section 5.4 can be specified by the weights of the assets. These weights may depend on both time and the asset prices. The optimal portfolios in this chapter can be considered as trading strategies with fixed weights (that do not depend on time or the asset prices).

These weights will change when the asset prices change. We therefore have to rebalance the portfolios from time to time. It turns out that the value of a frequently rebalanced portfolio is proportional to a geometric mean of the asset prices.

*Mathematical Asset Management.* By Thomas Höglund

## 6.1 RISK AND DIVERSIFICATION

Consider a portfolio consisting of $m$ assets. We shall sometimes call the assets stocks, but there are other possibilities such as bonds, forwards, and options. The asset prices at time $t$ are $S_1(t), ..., S_m(t)$. Let $V_j(t)$ denote the value of the holding of the asset $j$ at time $t$, $j = 1, ..., m$; $V_j(t) = a_j S_j(t)$, where $a_j$ is the number of asset $j$ in the portfolio. The portfolio value at time $t$, $P(t)$, is given by

$$P(t) = V_1(t) + \cdots + V_m(t), \tag{6.1}$$

and the asset $j$ has the weight

$$v_j(t) = V_j(t)/P(t)$$

in the portfolio. The return of the portfolio in the time interval $(t, t + \partial t)$, $R_P(t, t + \partial t)$, is given by

$$R_P(t, t + \partial t) = \frac{P(t + \partial t) - P(t)}{P(t)} = \sum_{j=1}^{m} v_j(t) R_j(t, t + \partial t), \tag{6.2}$$

where

$$R_j(t, t + \partial t) = \frac{V_j(t + \partial t) - V_j(t)}{V_j(t)} = \frac{S_j(t + \partial t) - S_j(t)}{S_j(t)} \tag{6.3}$$

is the return of the asset $j$ during the time interval.

EXERCISE 6.1. Show this.

The return of the portfolio has variance

$$\text{Var}(R_P(t, t + \partial t)) = \sum_{j=1}^{m} \sum_{k=1}^{m} v_j(t) v_k(t) \text{Cov}(R_j(t, t + \partial t), R_k(t, t + \partial t)) =$$

$$\boldsymbol{v}(t) \cdot Q_{\partial t} \boldsymbol{v}(t),$$

where $\boldsymbol{v}(t) = (v_1(t), ..., v_m(t))$. (Here and in the sequel we write vectors as row vectors, but they work as column vectors in matrix calculations.)

We have seen in Table 4.5 that $Q_{\partial t} \approx Q \partial t$ with good approximation, where $Q$ is the covariance matrix for the yearly growths of the stocks. It is therefore of no importance which of these matrices we use. Here we shall use

$$\sigma_P(t) = \sqrt{\boldsymbol{v}(t) \cdot Q \boldsymbol{v}(t)} \tag{6.4}$$

as a measure of the portfolio risk.

Assume that we have found that certain weights are optimal in some sense and that we form a portfolio with these weights. The weights will change with time; and when they differ essentially from the optimal, we have to rebalance the portfolio (decrease the holdings that have been too large, and increase those who have been too small) so that the weights are restored. We shall return to the rebalanced portfolio in Section 6.3. It turns out that the volatility of this portfolio coincides with $\sigma_P$ as defined in (6.4). We shall therefore call this quantity the *portfolio volatility*.

The risk can be decreased by diversification. Skanska (has the lowest volatility of the stocks in the example portfolio *Five Stocks* during the Period 1–4, $0.27$, whereas Ericsson has the highest, $0.49$. (See Table 4.1.) As an alternative to only have the "safe" stock Skanska, one can decrease the risk slightly by adding the "risky" stock Ericsson: Let $v$ be the weight of Ericsson and let $1 - v$ be the weight of Skanska. For this portfolio we have

$$\sigma_P^2 = 0.49^2 v^2 + 0.27^2 (1 - v)^2 + 2 \cdot 0.30 \cdot 0.49 \cdot 0.27 v (1 - v)$$

since the correlation equals $0.30$ according to Table 4.6.

This expression is minimized for $v = 0.14$, and the minimal volatility equals $0.26$.

EXERCISE 6.2. Carry through the details of the above reasoning.

AstraZeneca has higher volatility, $0.32$, than Skanska. In order to make the point more clear, we shall not use the correlation $0.20$ during Period 1–4 but $\rho = -0.07$, which is the lowest correlation between AZN and LME during the four subperiods. The portfolio that has the weight $0.68$ in AZN and $0.32$ in LME also has the volatility $0.26$. Thus, high volatility can be compensated by low covariation.

In order to get an idea of how the number of stocks, $m$, influence the risk, it is instructive to look at the case when all the weights are equal, $1/m$, all stocks have the same volatility, $\sigma$, and all correlations are the same, $\rho$. (This is possible for all $m$ if and only if $\rho \geq 0$.) In this case

$$\sigma_P = \sigma \sqrt{\rho + \frac{1 - \rho}{m}}. \tag{6.5}$$

EXERCISE 6.3. Show this, and show more generally that (6.5) holds with $\sigma$ replaced by the harmonic mean value of the volatilities, $\sigma_1, \ldots, \sigma_m$, in the case when the weights are inversely proportional to the volatilities, $v_i = C/\sigma_i$, and all correlations are the same.

The portfolio risk decreases to $\sigma\sqrt{\rho}$, as $m \to \infty$. If $\rho > 0$, which is the normal case, then there is a limit for what can be achieved by diversification of a stock portfolio. If one wants to reduce the risk further, one can add other assets such as bonds or real estates, or simply put (a part of) the money on the cash account. Another possibility is hedge funds (see Section 7.2.3) that can be uncorrelated or negatively correlated with the stock market.

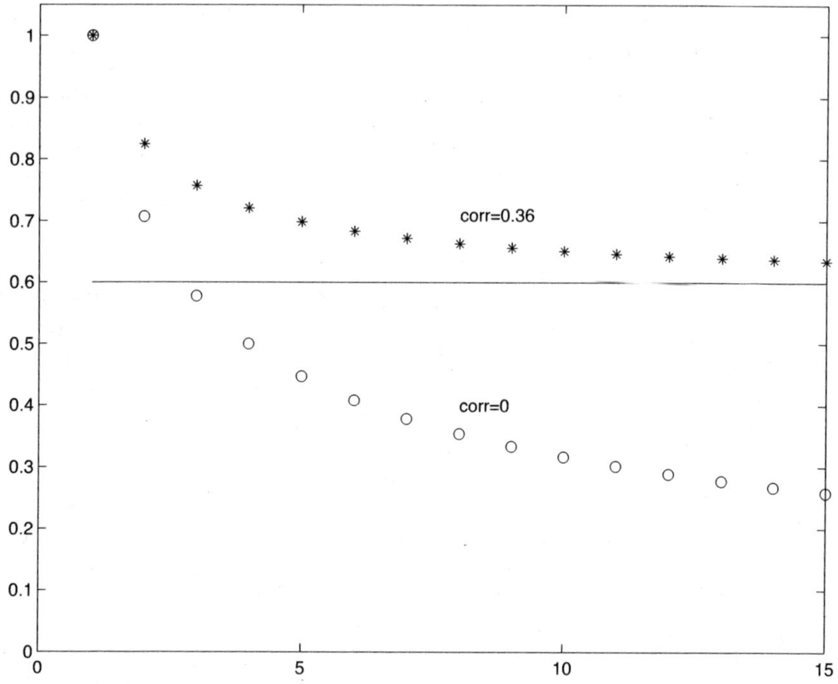


**Figure 6.1.** Portfolio volatility as a function of the number of assets.

In Fig. 6.1 the portfolio volatility is plotted as a function of $m$ in (the typical) case $\rho = 0.36$ and for $\rho = 0$. (The stock volatility equals 1 in the figure.) In the first case the portfolio volatility decreases from $\sigma$ to $0.6\sigma$, and already for $m = 5$ one has achieved 3/4 of this decrease, $\sigma_P = 0.7\sigma$.

### 6.1.1 The Minimum-Variance Portfolio

We shall determine the weights that minimizes the portfolio risk. This will sometimes give portfolios with negative weights. A negative holding is obtained if one borrows a stock and sells it (in order to buy back later).

The minimal variance equals zero if $Q$ is singular. Here we shall assume that this is not the case, but $Q$ is assumed to be nonsingular; that is, all the eigenvalues are strictly positive.

The portfolio with the lowest volatility is given by the weights $\boldsymbol{v}$ that minimizes

$$\frac{1}{2}\boldsymbol{v} \cdot Q\boldsymbol{v} = \frac{1}{2}\sum_i \sum_j v_i \sigma_{i,j} v_j$$

subject to the constraint $\sum_i v_i = 1$.

Using Lagrange multipliers, we get the equations

$$\sum_j \sigma_{i,j} v_j = \lambda,\ i = 1, ..., m,\ \sum_j v_j = 1. \tag{6.6}$$

EXERCISE 6.4. Determine the weights and the variance of the minimum-variance portfolio in the following three cases:

(a)

$$Q = \begin{pmatrix} \sigma_1^2 & 0 & 0 \\ 0 & \sigma_2^2 & 0 \\ 0 & 0 & \sigma_3^2 \end{pmatrix},$$

(b)

$$Q = \begin{pmatrix} 0.2 & 0.1 & 0.1 \\ 0.1 & 0.2 & 0.1 \\ 0.1 & 0.1 & 0.2 \end{pmatrix},$$

(c)

$$Q = \begin{pmatrix} 0.2 & 0 & 0 \\ 0 & 0.2 & 0.1 \\ 0 & 0.1 & 0.2 \end{pmatrix}.$$

The system (6.6) can also be written

$$Q\boldsymbol{v} = \lambda\mathbf{1},\ \mathbf{1} \cdot \boldsymbol{v} = 1, \tag{6.7}$$

where $\mathbf{1} = (1, ..., 1)$. We get

$$\boldsymbol{v} = \lambda Q^{-1}\mathbf{1}.$$

The constraint then gives

$$\lambda \mathbf{1} \cdot Q^{-1}\mathbf{1} = 1.$$

The minimal variance is

$$\boldsymbol{v} \cdot Q\boldsymbol{v} = \lambda^2 \mathbf{1} \cdot Q^{-1}\mathbf{1} = 1/\mathbf{1} \cdot Q^{-1}\mathbf{1}.$$

Note that

$$\mathbf{1} \cdot Q^{-1}\mathbf{1} > 0$$

since $Q$ and hence also $Q^{-1}$ is strictly positive definite.

**Summary 6.1** *The minimum-variance portfolio has variance*

$$\sigma^2_{min} = 1/\mathbf{1} \cdot Q^{-1}\mathbf{1} \tag{6.8}$$

*and the weights*

$$\boldsymbol{v}_{min} = \sigma^2_{min} Q^{-1}\mathbf{1}. \tag{6.9}$$

***Example 6.2.*** Uncorrelated Returns

In this case $\sigma_{i,i} = \sigma_i^2$ and $\sigma_{i,j} = 0$ for $i \neq j$. $Q^{-1}$ thus equals the diagonal matrix with the diagonal elements $1/\sigma_i^2$, $i = 1, ..., m$. We therefore have

$$\sigma^2_{min} = \frac{H}{m}, \; v_i = \frac{\sigma^2_{min}}{\sigma_i^2},$$

where $H$ denotes the harmonic mean value of $\sigma_1^2, ..., \sigma_m^2$,

$$H = \frac{m}{\frac{1}{\sigma_1^2} + \cdots + \frac{1}{\sigma_m^2}}.$$

It is seen here that the variance can be made arbitrarily small by diversifying the portfolio (choose $m$ large) in such a way that $H$ stays bounded. It is also seen that the minimum-variance portfolio has positive weights in this case.

EXERCISE 6.5. A capital is invested in a stock portfolio which has volatility 0.27. One considers two alternatives in order to decrease the risk: Half of the capital is moved from stocks to

(a) cash (which has volatility 0) and

(b) a certain hedge fund that has volatility 0.16 and the correlation $-0.50$ with the stock portfolio.

Which of the two alternatives will give the lowest volatility?

EXERCISE 6.6. A stock portfolio and a hedge fund have the correlation $\rho < 0$ and the volatilities $\sigma$ and $\tau$, respectively. Consider the portfolio that has the weight $p$ in the stock portfolio, the weight $w$ in the hedge fund, and the remainder, $1 - p - w$, in cash. Here $p$ is a given positive number. Determine $w$ so that the volatility of the portfolio is minimized, and calculate this minimal volatility.

Note that the volatility of the portfolio does not depend on the volatility of the hedge fund, but the weights do. Also note that the cash holding is negative if $p > \frac{1}{1+\kappa}$.

EXERCISE 6.7. Consider the same situation as in the above exercise with $p = 0.5$ and $\sigma = 0.27$. Calculate the weights and the volatility for the minimum-variance portfolio in the following two cases:

(a) $\tau = 0.16$ and $\rho = -0.50$.

(b) $\tau = 0.04$ and $\rho = -0.75$.

In the latter case the cash holding is negative. Assume that you do not accept this but instead choose the weight $0.5$ for the hedge fund in (b).

(c) Determine the volatility of this portfolio.

In Fig. 6.2 is a plot of the volatility of a portfolio that has the weight $v$, $0 \leq v \leq 1$, in a stock portfolio with volatility $0.27$ and also has the weight $1 - v$ in either cash or one of the two funds in Exercises 6.7a and 6.7b.

It is seen that as long as one has at least half of the weight in the stock portfolio, with the remainder in one of these assets, the largest decrease in volatility is achieved by investing in the most volatile of the alternatives.

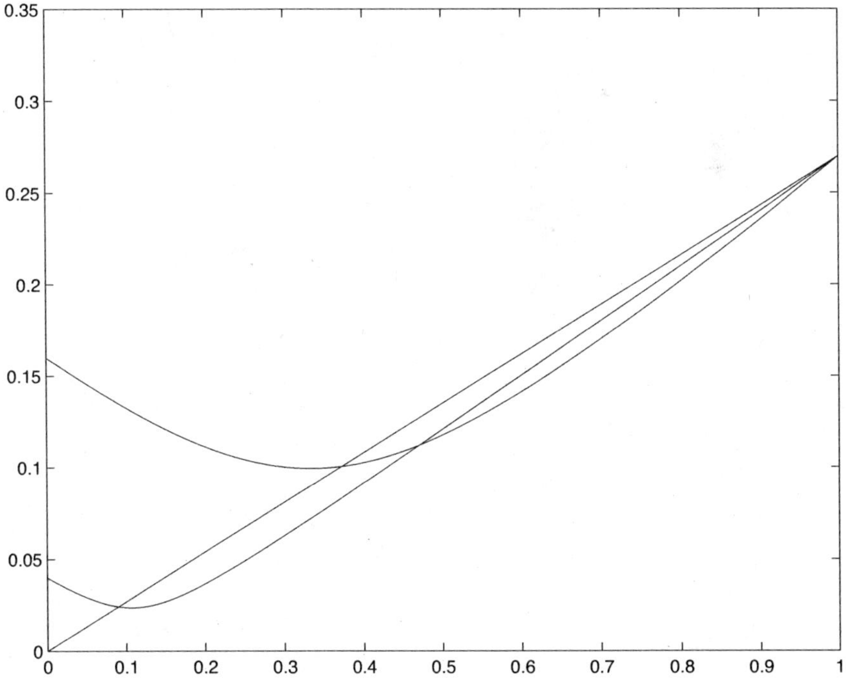


**Figure 6.2.** The volatility of the portfolio that consists of stocks and either of cash or one of the two funds in Exercise 6.7a and 6.7b.

### 6.1.2 Stability of the Estimates of the Weights

One way to estimate the weights of the minimum-variance portfolio is the following: Estimate $\boldsymbol{v}_{min}$ by

$$\hat{\boldsymbol{v}}_{min} = \hat{\sigma}^2_{min}\hat{Q}^{-1}\mathbf{1}, \tag{6.10}$$

where

$$\hat{\sigma}^2_{min} = 1/\mathbf{1}\cdot\hat{Q}^{-1}\mathbf{1}.$$

The estimate $\hat{Q}$ of $Q$ is based on historical data consisting of $n$ observations. Then use these weights for the minimum-variance portfolio during a period in the future. This portfolio is thus similar to the minimum-variance portfolio during the future period if the minimum-variance portfolios during the two periods are similar and if $n$ is so large that the estimate is stable.

***Example 6.3.*** *Five Stocks.*

The period October 1996—October 2000 is divided in four subperiods of length $n = 256$ trading days as in Chapter 4. The covariance matrix, $Q$, were estimated from the daily closing prices. The weights of the minimum-variance portfolio are given in Table 6.1.

**Table 6.1.** The weights of the minimum variance portfolio

| Period | AZN | LME | HM | SDIA | SKA |
|---|---|---|---|---|---|
| 1 | 0.20 | 0.06 | 0.16 | 0.11 | 0.47 |
| 2 | 0.42 | 0.01 | 0.19 | −0.14 | 0.53 |
| 3 | 0.33 | 0.01 | 0.19 | 0.09 | 0.38 |
| 4 | 0.31 | 0.09 | 0.07 | 0.02 | 0.51 |
| 1-2 | 0.35 | 0.01 | 0.17 | −0.05 | 0.52 |
| 3-4 | 0.32 | 0.07 | 0.10 | 0.04 | 0.47 |
| 1-4 | 0.34 | 0.05 | 0.12 | 0.00 | 0.48 |

A certain stability is seen here which becomes apparent if one ranks the weights as in Table 6.2. □

Table 6.3 shows the volatilities of the minimum-variance portfolios, along with the volatilities, $\sigma'_{min}$, of the portfolios that have the same weights as the minimum-variance portfolio during the preceding period. This is thus the volatility one gets using the above method. For a comparison, also the volatility of AFGX, $\sigma_{\text{afgx}}$, and the volatility of the portfolio that has equal weights of the five stocks, $\sigma_{\text{equal}}$, are given.

It is seen that it is not necessary to have many stocks in a portfolio to reduce the risk to the same level as AFGX.

**Table 6.2.** The ranks of the weights of the minimum variance portfolio

| Period | AZN | LME | HM | SDIA | SKA |
|---|---|---|---|---|---|
| 1 | 2 | 5 | 3 | 4 | 1 |
| 2 | 2 | 4 | 3 | 5 | 1 |
| 3 | 2 | 5 | 3 | 4 | 1 |
| 4 | 2 | 3 | 4 | 5 | 1 |
| 1-2 | 2 | 4 | 3 | 5 | 1 |
| 3-4 | 2 | 4 | 3 | 5 | 1 |
| 1-4 | 2 | 4 | 3 | 5 | 1 |

**Table 6.3.** Volatilities of three portfolios

| Period | $\sigma_{min}$ | $\sigma'_{min}$ | $\sigma_{\text{equal}}$ | $\sigma_{\text{afgx}}$ |
|---|---|---|---|---|
| 1 | 0.18 | – | 0.20 | 0.15 |
| 2 | 0.26 | 0.29 | 0.33 | 0.28 |
| 3 | 0.21 | 0.22 | 0.23 | 0.19 |
| 4 | 0.21 | 0.23 | 0.28 | 0.29 |
| 1–4 | 0.22 | – | 0.26 | 0.23 |

We shall use the next theorem to get an idea of the size of the random error of the estimate of the weights.

**Theorem 6.4** *Assume that the stock prices develop according to Model B in Section 4.3. Then (6.10) is, as $n \to \infty$, asymptotically normally distributed with expectation $\boldsymbol{v}_{min}$ and covariance matrix*

$$\frac{1}{n}(\sigma_{min}^2 Q^{-1} - \boldsymbol{v}_{min}\boldsymbol{v}_{min}^T). \tag{6.11}$$

Note that the covariance matrix is singular. It maps $\mathbf{1}$ into 0.

*Proof.* Let $\boldsymbol{Z}_n = \sqrt{n}(\hat{Q} - Q)$. Then by Theorem 4.4, $\boldsymbol{Z}_n$ converges in distribution to a normally distributed random matrix, $\boldsymbol{Z}$, having zero expectation, and

$$E[Z_{i,k}Z_{j,l}] = \sigma_{i,j}\sigma_{k,l} + \sigma_{i,l}\sigma_{k,j}.$$

We have

$$\hat{Q}^{-1} = (I + Q^{-1}(\hat{Q} - Q))^{-1}Q^{-1} = Q^{-1} - Q^{-1}\boldsymbol{Z}_n Q^{-1}/\sqrt{n} + O(1/n),$$

and hence by (6.10)

$$\hat{\boldsymbol{v}}_{min} = \frac{\hat{Q}^{-1}\mathbf{1}}{\mathbf{1}\cdot\hat{Q}^{-1}\mathbf{1}} = \boldsymbol{v}_{min} - \boldsymbol{u}_n/\sqrt{n} + O(1/n),$$

where

$$\boldsymbol{u}_n = Q^{-1}\boldsymbol{Z}_n\boldsymbol{v} - (\boldsymbol{1}\cdot Q^{-1}\boldsymbol{Z}_n\boldsymbol{v})\boldsymbol{v}.$$

Here and below $\boldsymbol{v} = \boldsymbol{v}_{min}$. The random vector $\boldsymbol{u}_n$ thus tends in distribution to the normally distributed random vector

$$\boldsymbol{u} = Q^{-1}\boldsymbol{w} - (\boldsymbol{1}\cdot Q^{-1}\boldsymbol{w})\boldsymbol{v}, \tag{6.12}$$

where $\boldsymbol{w} = \boldsymbol{Z}\boldsymbol{v}$. It therefore remains to calculate $\mathrm{Cov}(\boldsymbol{u})$.

We have

$$\boldsymbol{u}\boldsymbol{u}^T = Q^{-1}[\boldsymbol{w}\boldsymbol{w}^T - \boldsymbol{w}\boldsymbol{w}^T\boldsymbol{v}\boldsymbol{1}^T - \boldsymbol{1}\boldsymbol{v}^T\boldsymbol{w}\boldsymbol{w}^T + \boldsymbol{1}\boldsymbol{v}^T\boldsymbol{w}\boldsymbol{w}^T\boldsymbol{v}\boldsymbol{1}^T]Q^{-1} \tag{6.13}$$

and

$$E[\boldsymbol{w}\boldsymbol{w}^T] = \sigma^2_{min}Q + (Q\boldsymbol{v})(Q\boldsymbol{v})^T. \tag{6.14}$$

This is so because

$$E[(\boldsymbol{Z}\boldsymbol{v})_i(\boldsymbol{Z}\boldsymbol{v})_j] = \sum_k\sum_l E[Z_{i,k}Z_{j,l}]v_k v_l =$$

$$\sum_k\sum_l(\sigma_{i,j}\sigma_{k,l} + \sigma_{i,l}\sigma_{k,j})v_k v_l = \sigma_{i,j}\boldsymbol{v}\cdot Q\boldsymbol{v} + (Q\boldsymbol{v})_i(Q\boldsymbol{v})_j \tag{6.15}$$

and $\boldsymbol{v}\cdot Q\boldsymbol{v} = \sigma^2_{min}$.

It follows that

$$E[\boldsymbol{u}\boldsymbol{u}^T] = \sigma^2_{min}Q^{-1} + \boldsymbol{v}\boldsymbol{v}^T - 2\boldsymbol{v}\boldsymbol{v}^T - 2\boldsymbol{v}\boldsymbol{v}^T + 2\boldsymbol{v}\boldsymbol{v}^T = \sigma^2_{min}Q^{-1} - \boldsymbol{v}\boldsymbol{v}^T.$$

□

We shall use

$$d_{\text{theor}} = \sqrt{\frac{1}{m}E|\hat{\boldsymbol{v}}_{min} - \boldsymbol{v}_{min}|^2}$$

as a measure of the average deviation between $\hat{\boldsymbol{v}}_{min}$ and $\boldsymbol{v}_{min}$.

Assume that Model B holds and that $\hat{\boldsymbol{v}}_{min}$ is normally distributed with the same expectation and variance as the asymptotic distribution in the above theorem. Then

$$d_{\text{theor}} = \sqrt{\frac{1}{nm}\text{trace}(\sigma^2_{min}Q^{-1} - \boldsymbol{v}_{\min}\boldsymbol{v}_{min}{}^T)}.$$

Here trace stands for the sum of the diagonal elements of the matrix. We shall estimate $d_{\text{theor}}$ with

$$\hat{d}_{\text{theor}} = \sqrt{\frac{1}{nm}\frac{\text{trace}_1 + \cdots + \text{trace}_A}{A}},$$

where $A$ is the number of periods, $m$ is the common length of the periods, and trace$_t$ is the trace of the estimate of the covariance matrix based on data from period $t$, $t = 1, ..., A$.

The observed average deviation between the weights consecutive periods is

$$d'_{\text{obs}} = \sqrt{\frac{1}{A-1}\sum_{t=1}^{A-1}\frac{1}{m}|\hat{\boldsymbol{v}}_{min}(t+1) - \hat{\boldsymbol{v}}_{min}(t)|^2},$$

where $\hat{\boldsymbol{v}}_{min}(t)$ is the estimate of the weights during period $t$.

Model B implies that $\hat{\boldsymbol{v}}(1)$,...,$\hat{\boldsymbol{v}}(A)$ are independent, and hence

$$E(d'_{\text{obs}})^2 = \frac{1}{m}E|\hat{\boldsymbol{v}}_{min}(2) - \hat{\boldsymbol{v}}_{min}(1)|^2 = \frac{2}{m}E|\hat{\boldsymbol{v}}_{min}(1) - E\hat{\boldsymbol{v}}_{min}(1)|^2$$

and hence

$$d'_{\text{obs}} \approx \sqrt{2}d_{\text{theor}}.$$

In order to get comparable quantities that measures the distance between estimated and real weights, we shall therefore put

$$d_{\text{obs}} = d'_{\text{obs}}/\sqrt{2}.$$

These distances are given for a number of different period lengths in Table 6.4. The estimates are based on the daily growths, and the figures are rounded to nearest percent.

**Table 6.4.** Distances between the weights. Daily growths

| Period Length | Number of Periods | $\hat{d}_{\text{theor}}$ | $d_{\text{obs}}$ | $d_{\text{obs}}/\hat{d}_{\text{theor}}$ |
|---|---|---|---|---|
| 1024 | 1 | 0.02 | – | – |
| 512 | 2 | 0.03 | 0.04 | 1.52 |
| 256 | 4 | 0.04 | 0.09 | 2.22 |
| 128 | 8 | 0.06 | 0.11 | 1.90 |
| 64 | 16 | 0.08 | 0.11 | 1.43 |

It is seen that the theory is on the optimistic side. The reason might be that we have used the daily growths, and these are not normally distributed. Table 6.5 shows the corresponding figures when we replace the daily growths by the weekly (5 days) growths. The period length in the table is still measured in days.

**Table 6.5.** Distances between the weights. Weekly growths

| Period Length | Number of Periods | $\hat{d}_{\text{theor}}$ | $d_{\text{obs}}$ | $d_{\text{obs}}/\hat{d}_{\text{theor}}$ |
|---|---|---|---|---|
| 1020 | 1 | 0.04 | – | – |
| 510 | 2 | 0.06 | 0.05 | 0.85 |
| 255 | 4 | 0.08 | 0.09 | 1.04 |
| 125 | 8 | 0.12 | 0.13 | 1.05 |
| 65 | 15 | 0.20 | 0.18 | 0.88 |

It is seen that the agreement between theory and data has improved. Note that the observed precision is only slightly lower despite the fact that the number of observations is reduced by a factor 5.

One conceivable possibility is that the weights change with time, and therefore the best is to use fairly short periods of observation. This is thus *not* seen from the above table, but on the contrary the distances are monotonically decreasing functions of the length of the period of observation. Comparing the deviations with the mean weight $1/m = 20\%$ the conclusion becomes:

**Conclusion 6.5.** *Use, if possible, observations from several years.*

## 6.2 GROWTH PORTFOLIOS

One does not buy stocks in order to minimize the risk, but perhaps to maximize the growth. In this section we shall consider portfolios that consist of cash and $m$ stocks and will determine the portfolios that have maximal growth under various constraints.

It turns out that all these portfolios are linear combinations of three portfolios: cash, the minimum variance portfolio, and a certain auxiliary stock portfolio whose weights sum to zero.

Let $\boldsymbol{v} = (v_1, ..., v_m)$ and $\boldsymbol{\mu} = (\mu_1, ..., \mu_m)$ denote the weights and expected returns of the stocks in the portfolio. The expected return of the stock $i$, $\mu_i$, is defined as the instantaneous expected return of asset $i$ per unit time (for example, the expected daily return multiplied by 250).

The weight of cash then equals $1 - v_1 - \cdots - v_m$. The cash account yields $r_f$ per year. The expected return of the portfolio, $\mu_P$, therefore satisfies

$$\mu_P = r_f(1 - v_1 - \cdots - v_m) + \mu_1 v_1 + \cdots + \mu_m v_m = r_f + (\boldsymbol{\mu} - r_f\mathbf{1}) \cdot \boldsymbol{v}.$$

We shall assume (see Table 4.4) that the expected yearly growth (i.e., the yearly drift), $\nu_P$, of the portfolio is given by

$$\nu_P = \mu_P - \frac{1}{2}\sigma_P^2.$$

Here

$$\sigma_P^2 = \sum_1^m \sum_1^m \sigma_{i,j} v_i v_j = \boldsymbol{v} \cdot Q\boldsymbol{v}$$

is the portfolio variance per year. Thus

$$\nu_P = r_f + (\boldsymbol{\mu} - r_f \mathbf{1}) \cdot \boldsymbol{v} - \frac{1}{2}\boldsymbol{v} \cdot Q\boldsymbol{v}. \tag{6.16}$$

### 6.2.1 The Auxiliary Portfolio

This is the stock portfolio with the weights

$$\boldsymbol{v}_{aux} = Q^{-1}(\boldsymbol{\mu} - \mu_{min}\mathbf{1}). \tag{6.17}$$

Here $\mu_{min}$ is the expected return of the minimum-variance portfolio;

$$\mu_{min} = \boldsymbol{\mu} \cdot \boldsymbol{v}_{min} = \sigma_{min}^2 \mathbf{1} \cdot Q^{-1}\boldsymbol{\mu}, \tag{6.18}$$

and $\sigma_{min}^2$ is given by (6.8).

Define

$$\mu_{aux} = \boldsymbol{\mu} \cdot \boldsymbol{v}_{aux} \text{ and } \sigma_{aux}^2 = \boldsymbol{v}_{aux} \cdot Q\boldsymbol{v}_{aux}. \tag{6.19}$$

EXERCISE 6.8. Show that

$$\mu_{aux} = \sigma_{aux}^2 = (\boldsymbol{\mu} - \mu_{min}\mathbf{1}) \cdot Q^{-1}(\boldsymbol{\mu} - \mu_{min}\mathbf{1}). \tag{6.20}$$

Note that $\mu_{aux}$ does not stand for the expected return of the auxiliary portfolio. The return of this portfolio is not defined since its value at time 0 equals 0. We have, however, the following result.

EXERCISE 6.9. Show that the portfolio with the stock weights

$$a\boldsymbol{v}_{min} + b\boldsymbol{v}_{aux}, \tag{6.21}$$

and the weight $1 - a$ in cash has expected return

$$\mu_P = r_f + a\sigma_{min}^2\omega + b\sigma_{aux}^2, \tag{6.22}$$

variance

$$\sigma_P^2 = a^2\sigma_{min}^2 + b^2\sigma_{aux}^2, \tag{6.23}$$

and drift

$$\nu_P = r_f + \sigma_{min}^2 a(\omega - \frac{a}{2}) + \sigma_{aux}^2 b(1 - \frac{b}{2}). \tag{6.24}$$

Here

$$\omega = \frac{\mu_{min} - r_f}{\sigma_{min}^2}. \tag{6.25}$$

### 6.2.2 Maximal Drift

The drift (6.16) is a negative definite quadratic form, and hence it is maximized when

$$\boldsymbol{\mu} - r_f \mathbf{1} - Q\boldsymbol{v} = 0$$

that is, when $\boldsymbol{v} = \boldsymbol{v}_{max}$, where

$$\boldsymbol{v}_{max} = Q^{-1}(\boldsymbol{\mu} - r_f \mathbf{1}). \tag{6.26}$$

***Example 6.6.*** Uncorrelated Returns
In this case, $\boldsymbol{v}_{max}(i) = \frac{\mu_i - r_f}{\sigma_i^2}$ for $i = 1, ..., m$. □

The total weight of the stocks thus equals

$$\mathbf{1} \cdot \boldsymbol{v}_{max} = \mathbf{1} \cdot Q^{-1}(\boldsymbol{\mu} - r_f \mathbf{1}) = \omega.$$

EXERCISE 6.10. Show that the maximal drift portfolio has variance

$$\sigma_{max}^2 = (\boldsymbol{\mu} - r_f \mathbf{1}) \cdot Q^{-1}(\boldsymbol{\mu} - r_f \mathbf{1}) \tag{6.27}$$

and drift

$$\nu_{max} = r_f + \frac{1}{2}\sigma_{max}^2. \tag{6.28}$$

The weights can also be expressed on the form (6.21).

EXERCISE 6.11. Show that

$$\boldsymbol{v}_{max} = \omega \boldsymbol{v}_{min} + \boldsymbol{v}_{aux}. \tag{6.29}$$

It follows from (6.23) that

$$\sigma_{max}^2 = \omega^2 \sigma_{min}^2 + \sigma_{aux}^2. \tag{6.30}$$

In the case $\mu_{min} > r_f$, the stock portfolio given by the weights $\boldsymbol{v}_{max}/\omega$ is called the *Markowitz portfolio* or *Tangency portfolio*. (See Section 2.6 for an explanation of the names.) This portfolio consequently has the weights

$$\boldsymbol{v}_{Mz} = \sigma_{min}^2 \frac{Q^{-1}(\boldsymbol{\mu} - r_f \mathbf{1})}{\mu_{min} - r_f}. \tag{6.31}$$

EXERCISE 6.12. Show that the expected return, $\mu_{Mz}$, and the volatility, $\sigma_{Mz}$, of the Markowitz portfolio are given by

$$\mu_{Mz} = r_f + \sigma_{max}^2/\omega, \text{ and } \sigma_{Mz} = \sigma_{max}/\omega. \tag{6.32}$$

Maximal drift is thus achieved by investing the proportion $\omega$ in the Markowitz portfolio and keep the remainder in cash.

Note that it is possible to get higher growth than the stock market by keeping some of the money in cash even if the expected growth of the stock market is higher than the interest rate. The reason is that one moves money from stocks to cash when the value of the stock portfolio has increased more than the cash — and conversely. One thus uses the volatility.

### 6.2.3 Constraint on Portfolio Volatility

The risk can be decreased by imposing constraints on the portfolio volatility. That is, to maximize $\nu_P$ under the constraint $\sigma_P^2 = \sigma^2$. Let

$$\boldsymbol{x} = Q^{\frac{1}{2}}\boldsymbol{v} \text{ and } \boldsymbol{a} = Q^{-\frac{1}{2}}(\boldsymbol{\mu} - r_f\mathbf{1}). \tag{6.33}$$

We shall then maximize $\boldsymbol{a} \cdot \boldsymbol{x}$ under the constraint $|\boldsymbol{x}| = \sigma$. The maximum is attained when $\boldsymbol{x}$ has the same direction as $\boldsymbol{a}$ and hence $\boldsymbol{x} = (\sigma/|\boldsymbol{a}|)\boldsymbol{a}$.

The stock weights therefore satisfies

$$\boldsymbol{v} = \frac{\sigma}{\sigma_{max}}\boldsymbol{v}_{max}. \tag{6.34}$$

EXERCISE 6.13. Show that the drift of the above portfolio equals

$$r_f + \sigma_{max}\sigma - \frac{\sigma^2}{2} = \nu_{max} - \frac{1}{2}(\sigma - \sigma_{max})^2. \tag{6.35}$$

The portfolio weights (6.34) depend on the portfolio volatility $\sigma$, the interest rate $r_f$, the covariance matrix $Q$, and the expected returns $\boldsymbol{\mu}$. The volatility is chosen by us, the interest rate is known, and we have seen in Chapter 4 that $Q$ can be estimated with reasonable precision, and hence $Q$ can be considered as being known. The problem is the expected returns. They cannot be estimated with sufficient precision using historical data (Section 4.2). However, we do not have to be able to estimate the expected returns, but just the weights.

Assume that our estimate of $\boldsymbol{\mu}$ is $\boldsymbol{\mu}'$. Then our estimate of $\boldsymbol{v}_{max}$ becomes

$$\boldsymbol{v}' = Q^{-1}(\boldsymbol{\mu}' - r_f\mathbf{1})/\sigma'_{max}.$$

Here $\sigma'_{max}$ is as in (6.27), but with $\boldsymbol{\mu}$ replaced by $\boldsymbol{\mu}'$.

EXERCISE 6.14. Show that $\boldsymbol{v} = \boldsymbol{v}'$ if and only if there is a constant $a > 0$ such that $\mu'_i - r_f = a(\mu_i - r_f)$ for $i = 1, ..., m$.

Anticipating a result in Chapter 7, the CAPMidentity says that $\mu_i - r_f = \beta_i(\mu_M - r_f)$ for $i = 1, ..., m$. Here the parameters $\beta_1, ..., \beta_m$ can be estimated from historical data, and the above exercise shows that it is not necessary to be able to estimate $\mu_M$ (the expected return of the market portfolio). The weights (6.34) can therefore be estimated with reasonable precision if the CAPMidentity holds, and $\mu_M > r_f$.

In Fig. 6.3 the maximal drift is plotted as a function of the portfolio volatility. This figure is based on the portfolio *Five Stocks* using data from Period 1– 4. The expected returns were estimated using the CAPM-identity with $r_f = 0.04$. In this case, $\boldsymbol{v}_{max} = (0.31, 0.93, 0.32, 0.38, 0.19)$, and $\omega = 2.13$.

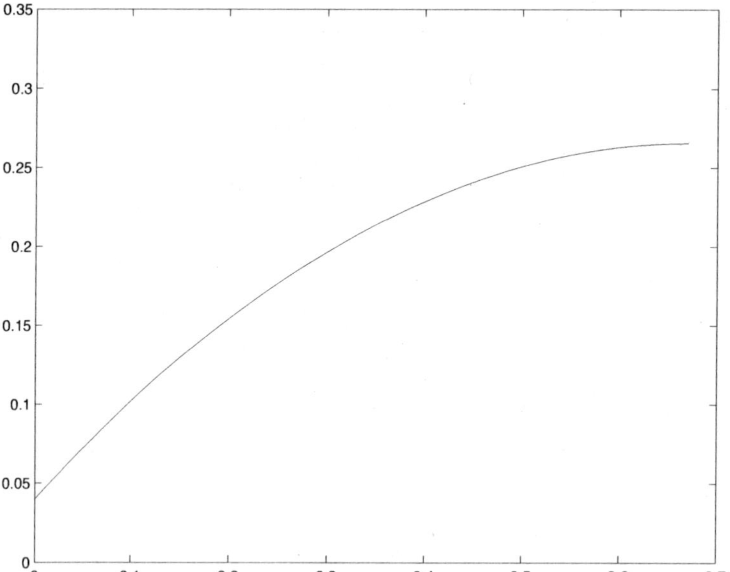


**Figure 6.3.** The maximal drift as a function of the portfolio volatility.

### 6.2.4 Constraints on Total Stock Weight

The amount of cash can be negative — it is possible to borrow money using the stocks as collateral. Normally the maximal pledge for stocks on the A-list on Stockholmsbörsen is 70%, and on the O-list it is 50% (and 90% for bonds) of their values. If one believes in a strong development of the stock market and wants maximal growth, one should therefore use maximal pledge.

Let $c_i$ be $0.7$ or $0.5$ depending on which list stock $i$ belongs to, and put $\mathbf{c} = (c_1, ..., c_m)$. Assume that the total weight of the stock portfolio is $h$, and hence the cash equals $-(h-1)$. The maximal pledge of the portfolio is $\mathbf{c} \cdot \boldsymbol{v}$. This gives the restriction

$$h \leq 1 + \mathbf{c} \cdot \boldsymbol{v} \tag{6.36}$$

Therefore it is not possible to construct the maximal drift portfolio if

$$\omega > 1 + \mathbf{c} \cdot \boldsymbol{v}_{max}. \tag{6.37}$$

For the portfolio in Fig. 6.3 we have $\omega = 2.13$, and the right-hand side of (6.37) equals $2.43$. This portfolio is thus possible to construct, but the reason for this is that most of the stocks belongs to the A-list. If all had belonged to the O-list, the right-hand side of (6.37) had equaled $2.07$, and hence this portfolio had not been possible.

There can be other reasons to put limitations on the total stock weight. One is that for reasons of risk we do not want to pledge the portfolio. Another is that because of regulations we are not allowed to do it. In these two cases the maximal total stock weight equals 1.

This leads to the problem to maximize the drift for a fixed total stock weight in the portfolio.

We shall thus maximize (6.16) under the constraint

$$\boldsymbol{v} \cdot \mathbf{1} = h.$$

Using Lagrange multipliers, we get the equations

$$\boldsymbol{\mu} - r_f \mathbf{1} - Q\boldsymbol{v} - \lambda \mathbf{1} = 0 \quad \text{and} \quad \mathbf{1} \cdot \boldsymbol{v} = h$$

That is,

$$\boldsymbol{v} = Q^{-1}(\boldsymbol{\mu} - r_f \mathbf{1} - \lambda \mathbf{1}) \quad \text{and} \quad \lambda = \sigma_{min}^2(\omega - h).$$

Therefore the stock weights are

$$h\boldsymbol{v}_{min} + \boldsymbol{v}_{aux}, \tag{6.38}$$

and the amount of cash equals $1 - h$. It is seen here that $h = 0$ does not imply that the portfolio consists of cash only, but that the stock weights sum to zero. Some of the stocks thus have negative weights.

The maximal pledge for this portfolio equals

$$h\mathbf{c} \cdot \boldsymbol{v}_{min} + \mathbf{c} \cdot \boldsymbol{v}_{aux}.$$

The debt $h - 1$ must not exceed this value, which gives the bound

$$h \leq \frac{1 + \boldsymbol{c} \cdot \boldsymbol{v}_{aux}}{1 - \boldsymbol{c} \cdot \boldsymbol{v}_{min}}. \tag{6.39}$$

Note that the bound (6.39) equals $1/(1-c)$ when $c_i = c$ for all $i$.

### 6.2.5 Constraints on Total Stock Weight and Volatility

Next consider the portfolio that has maximal drift under constraints on the total stock weight, and the portfolio volatility. This is the case for example if we want a fully invested stock portfolio ($h = 1$) under controlled risk.

We shall thus determine the weights $\boldsymbol{v}$ that maximize the drift under the constraints

$$\boldsymbol{v} \cdot \mathbf{1} = h \quad \text{and} \quad \boldsymbol{v} \cdot Q\boldsymbol{v} = \sigma^2.$$

Let $\boldsymbol{x}$ and $\boldsymbol{a}$ be as in (6.33), and let $\boldsymbol{b} = Q^{-\frac{1}{2}}\mathbf{1}$. We shall then maximize

$$\boldsymbol{a} \cdot \boldsymbol{x} \text{ subject to } |\boldsymbol{x}| = \sigma \quad \text{and} \quad \boldsymbol{b} \cdot \boldsymbol{x} = h.$$

Let $\boldsymbol{y} = \boldsymbol{x} - h\boldsymbol{b}/|\boldsymbol{b}|^2$. Then

$$\boldsymbol{y} \in \boldsymbol{b}^{\perp} = \{\boldsymbol{z} \in \mathbb{R}; \boldsymbol{b} \cdot \boldsymbol{z} = 0\}$$

and

$$|\boldsymbol{y}|^2 = |\boldsymbol{x}|^2 - \frac{h^2}{|\boldsymbol{b}|^2} = \sigma^2 - h^2\sigma_{min}^2. \tag{6.40}$$

We have $\boldsymbol{a} \cdot \boldsymbol{x} = h\boldsymbol{a} \cdot \boldsymbol{b}/|\boldsymbol{b}|^2 + \boldsymbol{a} \cdot \boldsymbol{y}$. Let $\boldsymbol{d} = \boldsymbol{a} - (\boldsymbol{a} \cdot \boldsymbol{b}/|\boldsymbol{b}|^2)\boldsymbol{b}$. Then $\boldsymbol{d} \in \boldsymbol{b}^{\perp}$ and $\boldsymbol{a} \cdot \boldsymbol{y} = \boldsymbol{d} \cdot \boldsymbol{y}$. We shall therefore maximize $\boldsymbol{d} \cdot \boldsymbol{y}$ subject to the constraints $\boldsymbol{y} \in \boldsymbol{b}^{\perp}$ and (6.40). Thus $\boldsymbol{y} = k\boldsymbol{d}$, where $k > 0$ and $k^2|\boldsymbol{d}|^2 = \sigma^2 - h^2\sigma_{min}^2$. Therefore $\boldsymbol{x} = h\boldsymbol{b}/|\boldsymbol{b}|^2 + k\boldsymbol{d}$.

EXERCISE 6.15. Show that $|\boldsymbol{d}|^2 = \sigma_{aux}^2$, and $\boldsymbol{a} \cdot \boldsymbol{b} = \omega$. Here $\sigma_{aux}$ and $\omega$ are as in (6.20) and (6.25).

The portfolio that has maximal drift under the constraints that it has volatility $\sigma$ and that the total stock weight equals $h$ therefore has the stock weights

$$\boldsymbol{v} = h\boldsymbol{v}_{min} + k\boldsymbol{v}_{aux}, \tag{6.41}$$

and the weight of cash equals $1 - h$. Here

$$k = \frac{\sqrt{\sigma^2 - h^2\sigma_{min}^2}}{\sigma_{aux}}.$$

### 6.2.6 The Efficient Frontier

The efficient frontier was introduced by Markowitz in his pioneering works 1952 and 1959. We shall not use it, but instead we shall derive the optimal portfolios directly without making any reference to the efficient frontier.

Instead of maximizing the expected *growth*, $\nu_P$, for a given portfolio volatility, $\sigma$, one can maximize the expected *return*, $\mu_P$, for a given portfolio volatility. The two problems have the same solution since $\mu_P = \nu_P + \sigma_P^2/2$, and it follows from Exercise 6.13 that the maximal expected return equals

$$r = r_f + \sigma_{max}\sigma. \tag{6.42}$$

The parabola (6.35) in the $(\sigma, g)$-plane thus corresponds to the straight line (6.42) in the $(\sigma, r)$-plane. This straight line is called the *efficient frontier*. It is thus possible to obtain a very high expected return by choosing $\sigma$ large, but the growth will be poor if $\sigma$ is considerably larger than $\sigma_{max}$.

Consider now a fully invested stock portfolio (no cash). We shall maximize the expected return of this portfolio for a given portfolio volatility. That is, maximize the drift for a given volatility, and total stock weight 1. This was done in the previous section. The weights are therefore given by (6.41) with $h = 1$.

EXERCISE 6.16. Use (6.41), (6.20), and the identity $r = \boldsymbol{\mu} \cdot \boldsymbol{v}$ to show that $k = \frac{r - \mu_{min}}{\sigma_{aux}^2}$

It follows that

$$\sigma^2 = \sigma_{min}^2 + \frac{(r - \mu_{min})^2}{\sigma_{aux}^2} \tag{6.43}$$

and that the weights are given by

$$\boldsymbol{v}_{min} + \frac{r - \mu_{min}}{\sigma_{aux}^2}\boldsymbol{v}_{aux}. \tag{6.44}$$

Not only the line (6.42), but also the part of the hyperbola (6.43) in the $(\sigma, r)$ plane for which $r \geq \mu_{min}$ and $\sigma > 0$ is called the efficient frontier.

EXERCISE 6.17. Show that if $r_f < \mu_{min}$, then the line (6.42) is the line through the point $(0, r_f)$ which is tangent to the curve (6.43). Also show that the weights (6.44) equals $\boldsymbol{v}_{Mz}$ at the tangent point.

This exercise explains the name tangency portfolio.

### 6.2.7 Summary

All portfolios we have considered have stock weights of the form (6.21) and the weight $1 - a$ in cash.

The constants $a$ and $b$ are given in Table 6.6. The expected returns and variances of the portfolios are given by (6.22) and (6.23).

**Table 6.6.** Weights of optimal portfolios

| Portfolio | $a$ | $b$ |
|---|---|---|
| Minimum variance | 1 | 0 |
| Maximal drift | $\omega$ | 1 |
| Maximal drift with volatility $\sigma$ | $\omega\sigma/\sigma_{max}$ | $\sigma/\sigma_{max}$ |
| Efficient frontier with cash | $\omega\sigma/\sigma_{max}$ | $\sigma/\sigma_{max}$ |
| Maximal drift with total stock weight $h$ | $h$ | 1 |
| Maximal drift with total stock weight $h$ and volatility $\sigma$ | $h$ | $\sqrt{\sigma^2 - h^2\sigma_{min}^2}/\sigma_{aux}$ |
| Efficient frontier | 1 | $\sqrt{\sigma^2 - \sigma_{min}^2}/\sigma_{aux}$ |

## 6.3 REBALANCING

The stock weights will change if the stock prices develop in different ways. In order to keep the weights, it is therefore necessary to rebalance the portfolio from time to time.

EXERCISE 6.18.

(a) Today the prices of three stocks are $0.50$, $16.85$, and $2.11$ EUR. Form a portfolio worth $10,000$ EUR and in which the three stocks have the weights $0.20$, $0.35$, and $0.45$, respectively. (The round off error is put on an account which is assumed to give the interest rate $0$.) How many of the different stocks do you need to buy?

(b) Assume that the portfolio is left unchanged until the stock prices are $0.34$, $20.00$, and $2.05$, respectively. Determine the weights of the stocks in the portfolio. How many of the different stocks do you have to buy or sell in order to restore the original weights?

If the portfolio is rebalanced at $t_0 < t_1 < ...$, then the value of the portfolio at $t_n$ equals

$$P(t_n) = P(t_0)\Pi_{k=1}^{n}\big(1 + R_P(t_{k-1}, t_k)\big), \tag{6.45}$$

where

$$R_P(t_{k-1}, t_k) = \frac{P(t_k) - P(t_{k-1})}{P(t_{k-1})} =$$

$$\sum_{i=1}^{m} \frac{V_i(t_k-) - V_i(t_{k-1}+)}{P(t_{k-1})} = \sum_{i=1}^{m} v_i R_i(t_{k-1}, t_k) \tag{6.46}$$

and where

$$R_i(t_{k-1}, t_k) = \frac{S_i(t_k) - S_i(t_{k-1})}{S_i(t_{k-1})}.$$

The stock holdings will have discontinuities at the moments of rebalancing. Therefore we have the right and left limits in (6.46).

EXERCISE 6.19. Carry through the details of the above reasoning.

### 6.3.1 The Portfolio as a Function of the Stocks

Let $v_1, ..., v_m$ be given weights. Consider the portfolio that has these weights from the beginning, and which is rebalanced at $0, \partial t, 2\partial t, 3\partial t, ...$ in such a way that the weights are restored. The value of the portfolio as a function of the values of the stocks is given by $G(t, \boldsymbol{S}(t))$, where $G$ is as in (5.36), provided that $\partial t$ is small and the stocks develop according to Model B of Chapter 4.

We shall here give a direct derivation of this result without making any reference to Section 5.4.

Let $n\partial t = t$, and let $P_n(t)$ denote the value of the portfolio at time $t$. It follows from (6.45) that

$$P_n(t) = P(0) \prod_{i=1}^{n} \sum_{j=1}^{m} v_j \frac{S_j(i\partial t)}{S_j((i-1)\partial t)}.$$

Also

$$\frac{S_j(i\partial t)}{S_j((i-1)\partial t)} = e^{\nu_j \partial t + \partial_i X_j},$$

where

$$\partial_i X_j = X_j(i\partial t) - X_j((i-1)\partial t) = \sqrt{\partial t} Z_j(i)$$

and where $\boldsymbol{Z}(i) = (Z_1(i), ..., Z_m(i))$, $i = 1, ..., n$ are independent, normally distributed random variables with zero expectation and covariance matrix $Q$.

Therefore

$$e^{\nu_j \partial t + \partial_i X_j} = 1 + \nu_j \partial t + \sqrt{\partial t} Z_j(i) + \frac{\partial t}{2} Z_j(i)^2 + O(\sqrt{\partial t}^3) =$$

$$1 + (\nu_j + \frac{1}{2}\sigma_{j,j})\partial t + \sqrt{\partial t} Z_j(i) + \frac{\partial t}{2} e_j(i) + O(\sqrt{\partial t}^3),$$

where

$$e_j(i) = Z_j(i)^2 - \sigma_{j,j}$$

and $\boldsymbol{e}(i) = (e_1(i), ..., e_m(i))$, $i = 1, ..., n$ are independent, identically distributed random variables with expectation 0, and $E|\boldsymbol{e}(i)|^2 < \infty$.

It follows that

$$\sum_{j=1}^{m} v_j \frac{S_j(i\partial t)}{S_j((i-1)\partial t)} = 1 + (\boldsymbol{v}\cdot\boldsymbol{\nu} + \frac{1}{2}\boldsymbol{v}\cdot\boldsymbol{d})\partial t + \sqrt{\partial t}\boldsymbol{v}\cdot\boldsymbol{Z}(i) + \frac{\partial t}{2}\boldsymbol{v}\cdot\boldsymbol{e}(i) + O(\sqrt{\partial t}^3),$$

where $\boldsymbol{\nu} = (\nu_1, ..., \nu_m)$ and $\boldsymbol{d} = (\sigma_{1,1}, ..., \sigma_{m,m})$. Therefore also

$$\ln(P_n(t)/P(0)) =$$

$$\sum_{i=1}^{n}((\boldsymbol{v}\cdot\boldsymbol{\nu} + \frac{1}{2}\boldsymbol{v}\cdot\boldsymbol{d})\partial t + \sqrt{\partial t}\boldsymbol{v}\cdot\boldsymbol{Z}(i) + \frac{\partial t}{2}\boldsymbol{v}\cdot\boldsymbol{e}(i) - \frac{1}{2}\partial t(\boldsymbol{v}\cdot\boldsymbol{Z}(i))^2 + O(\sqrt{\partial t}^3)) =$$

$$t\boldsymbol{v}\cdot\boldsymbol{\nu} + \boldsymbol{v}\cdot(\boldsymbol{X}(t) - \boldsymbol{X}(0)) + t\frac{1}{2}\boldsymbol{v}\cdot\boldsymbol{d} + \frac{\partial t}{2}\sum_{i=1}^{n}\boldsymbol{v}\cdot\boldsymbol{e}(i) - \frac{1}{2}\partial t\sum_{i=1}^{n}(\boldsymbol{v}\cdot\boldsymbol{Z}(i))^2 + O(\sqrt{\partial t}).$$

The random variable

$$\frac{\partial t}{2}\sum_{i=1}^{n}\boldsymbol{v}\cdot\boldsymbol{e}(i)$$

has expectation 0 and variance

$$(\partial t)^2 nE(\boldsymbol{v}\cdot\boldsymbol{e}(1))^2/4 = O(\partial t)$$

and therefore tends to zero in probability as $\partial t \to 0$.

The variable

$$\frac{1}{2}\partial t\sum_{i=1}^{n}(\boldsymbol{v}\cdot\boldsymbol{Z}(i))^2$$

has variance

$$(\partial t)^2 n\text{Var}((\boldsymbol{v}\cdot\boldsymbol{Z}(1))^2)/4 = O(\partial t)$$

and therefore converges to its expectation

$$t\frac{1}{2}\boldsymbol{v}\cdot Q\boldsymbol{v}.$$

We have thus proved the first part of the following theorem.

**Theorem 6.7.** *If the stocks develop according to Model B, then $P_n(t) \to P_\infty(t)$ in probability as $\partial t \to 0$. Here*

$$P_\infty(t) = P(0)e^{tL}(\frac{S_1(t)}{S_1(0)})^{v_1} \times \cdots \times (\frac{S_m(t)}{S_m(0)})^{v_m} \tag{6.47}$$

*and*

$$L = \frac{1}{2}(\sum_{j=1}^{m} v_j \sigma_{j,j} - \boldsymbol{v} \cdot Q\boldsymbol{v}). \tag{6.48}$$

*In particular:* $\ln(P_\infty(t)/P(0))$ *is normally distributed with expectation* $(\boldsymbol{v} \cdot \boldsymbol{\mu} - \frac{1}{2}\boldsymbol{v} \cdot Q\boldsymbol{v})t$ *and variance* $\boldsymbol{v} \cdot Q\boldsymbol{v}t$. *Here* $\mu_j = \frac{\sigma_{j,j}}{2} + \nu_j$ *are the expected returns.*

The statement about the distribution follows from the identity

$$\ln(P_\infty(t)/P(0)) = (\boldsymbol{v} \cdot \boldsymbol{\mu} - \frac{1}{2}\boldsymbol{v} \cdot Q\boldsymbol{v})t + \boldsymbol{v} \cdot \boldsymbol{X}(t). \tag{6.49}$$

Note that the theorem also holds when $Q$ is singular, which is the case for a portfolio with cash if the interest rate is constant.

We have the following corollary to Theorem 6.7.

**Corollary 6.8.** *Let $P_\infty(t)$ denote the value at time $t$ of the continuously rebalanced portfolio with weights as in (6.21). If Model B holds, then $\ln(P_\infty(t)/P_\infty(0))$ is normally distributed with expectation and variance $t$ times (6.24) and (6.23), respectively.*

The value of the continuously rebalanced portfolio thus equals the geometric mean value of the stocks multiplied by $e^{tL}$. This portfolio value can be compared to the value of the unchanged portfolio

$$P_1(t) = P(0)\big(v_1\frac{S_1(t)}{S_1(0)} + \cdots + v_m\frac{S_m(t)}{S_m(0)}\big),$$

which is the arithmetic mean value.

The geometric mean never exceeds the arithmetic mean. Therefore

$$e^{-Lt}P_\infty(t) \leq P_1(t). \tag{6.50}$$

The minimum variance portfolio based on *Five Stocks* during Period 1– 4 has the value $L = 0.028$, whereas the portfolio with equal weights has $L = 0.048$.

### 6.3.2 Empirical Verification

Model B agrees with data fairly well when $t$ is large, but deviates significantly when $t$ is small, which it is in the proof of Theorem 6.7. It is therefore not immediately clear that (6.47) gives the value of a frequently rebalanced portfolio. However, here we shall verify the result empirically.

We shall consider three different portfolios which consist of the *Five Stocks*: the portfolio with equal weights, the minimum variance portfolio, and the market neutral portfolio that has minimal variance. By a market neutral portfolio we mean a portfolio whose returns are uncorrelated with the returns of the market. In this case the market is Affärsväldens generalindex, an index that covers about 300 Swedish stocks.

The first 769 days (Period 1– 3) will be used to estimate the weights of the different portfolios, and the remaining 256 days will be used to check if the portfolio which is rebalanced every $r$ days converges to the portfolio (6.47) as $r \to 0$.

We shall assume that the starting values of all portfolios are 1. For $i = 1, ..., 256$, let $\Pi_0(i)$ denote the value of the portfolio (6.47) with $Q = \hat{Q}$ after day $i$, and let $\Pi_r(i)$ be the value of the portfolio which is rebalanced every $r$ days, $r = 1, 2, ....$

There is little need to rebalance if the weights change only marginally. The effect of rebalancing is thus best illustrated by choosing a period during which the weights change considerably. The first 100 of the last 256 days is such a period. Ericsson and Skandia went up about $140\%$, Hennes $55\%$, and AstraZeneca and Skanska (were in comparison almost unchanged.

We shall measure the distances

$$d(r) = \frac{\sum_{i=1}^{100} |\Pi_r(i) - \Pi_0(i)|}{100}, \quad r = 1, 2, ...$$

Figure 6.4 shows the values of $d(r)$ for $r = 1, 2, ..., 100$ days for the portfolio with equal weights. Figure 6.5 shows the development of the portfolios $\Pi_0$, $\Pi_1$, and $\Pi_{100}$ during the first 100 days. The latter portfolio is thus never rebalanced.

Figures 6.6 to 6.9 are the corresponding plots for the minimum variance portfolio and the market neutral minimum variance portfolio. The weights of these portfolios are $(0.35, 0.01, 0.17, -0.01, 0.47)$, $(0.60, -0.58, 0.27, -0.20, 1.04)$, respectively.

The minimum variance portfolio has the heavy weights $0.35$ and $0.47$ in the low-volatility stocks AstraZeneca and Skanska(, whereas the high volatility stocks Ericsson and Skandia have the low weights $0.01$ and $-0.01$ of opposite signs. Therefore one may expect that there is little need of rebalancing. That it is so is seen in Fig. 6.6.

The reason to rebalance the minimum variance portfolio is that the variance becomes too big. An alternative to daily rebalancings is thus to wait to the first moment,

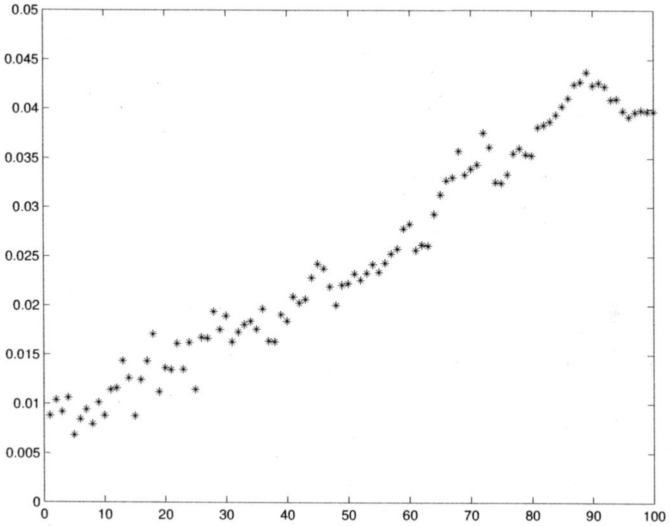


**Figure 6.4.** Equal weights. Distance between the portfolios that are rebalanced continuously and every $k$ days, as a function of $k$.

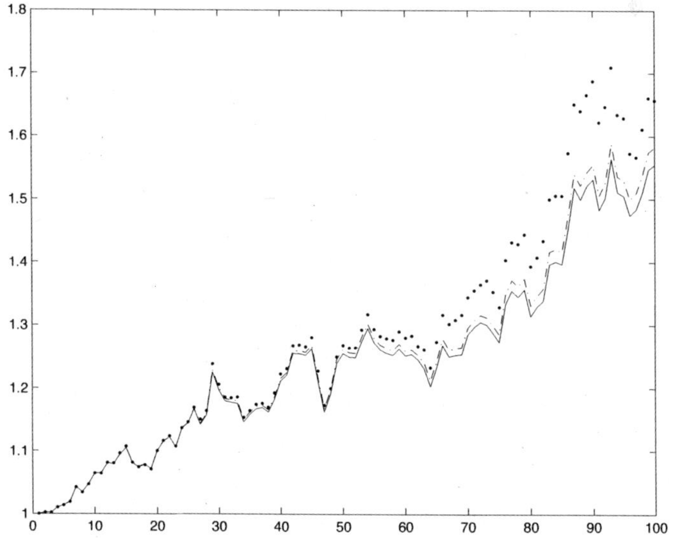


**Figure 6.5.** Equal weights. Development of the portfolios that are rebalanced continuously (—), daily (– · ), and never (· ).

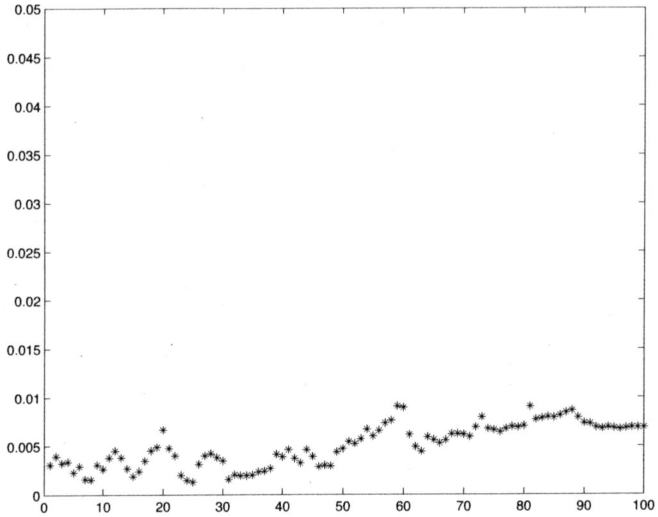


**Figure 6.6.** Minimum variance. Distance between the portfolios that are rebalanced continuously and every $k$ days, as a function of $k$.

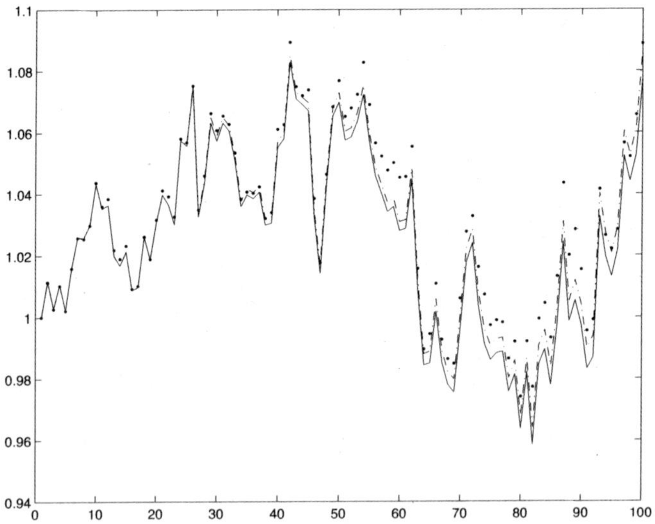


**Figure 6.7.** Minimum variance. Development of the portfolios that are rebalanced continuously (—), daily (– · ), and never (· ).

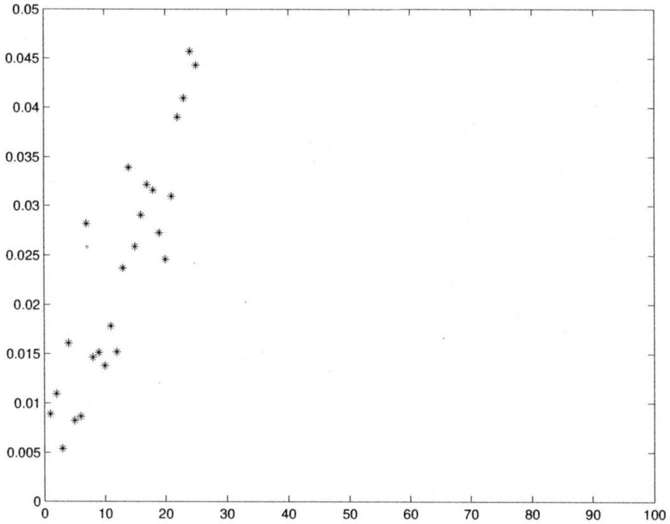


**Figure 6.8.** Market neutral. Distance between the portfolios that are rebalanced continuously and every $k$ days, as a function of $k$.

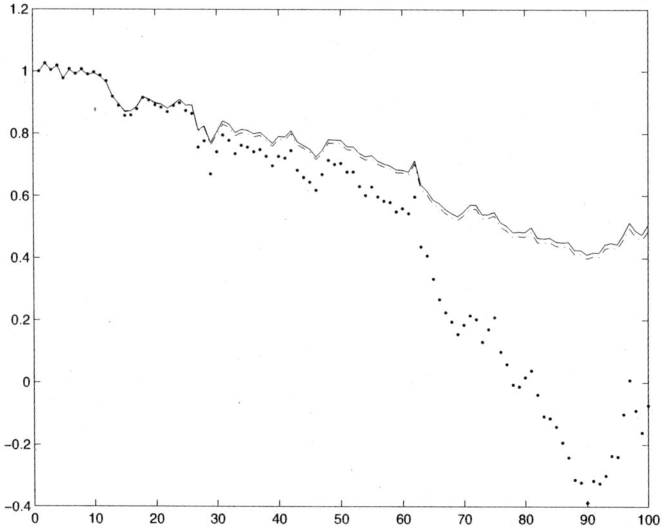


**Figure 6.9.** Market neutral. Development of the portfolios that are rebalanced continuously (—), daily (– · ), and never (· ).

$t$, at which $\sigma(t) \geq \sigma_{min}(1+\epsilon)$, where $\sigma(t) = \sqrt{\boldsymbol{v}(t) \cdot Q\boldsymbol{v}(t)}$ and where $\epsilon$ is an appropriately chosen positive number.

In Fig. 6.10 the quotient $\sigma(t)/\sigma_{min}$ is plotted for the unchanged minimum variance portfolio based on *Five Stocks*. The time period is Period 4, and the weights and $\sigma_{min}$ are estimated with data from Period 1– 3.

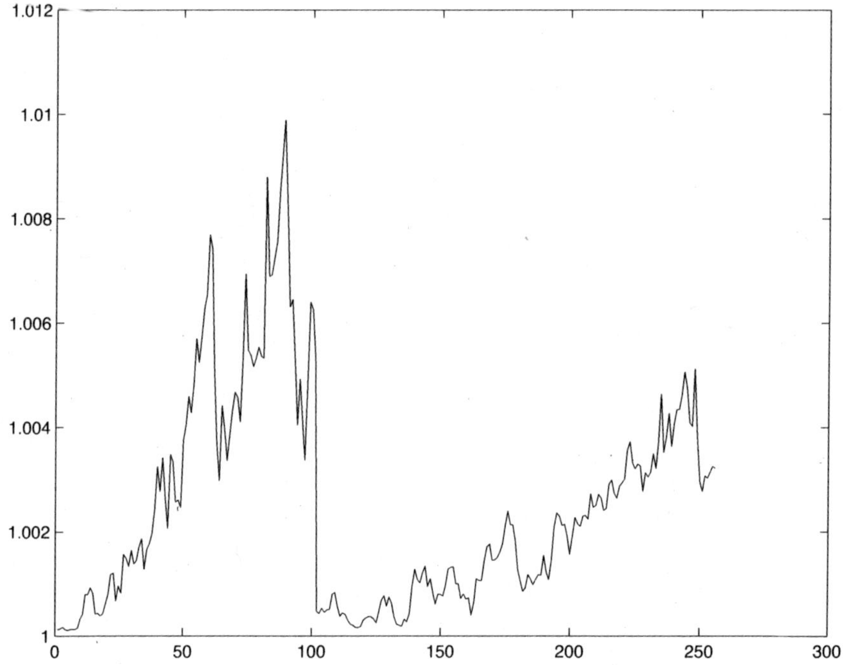


**Figure 6.10.** Portfolio volatility relative to the volatility of the minimum variance portfolio.

The quotient never exceeds $1.01$. For this reason there is little need to rebalance.

The quotients $d(1)/d(100)$ are $0.22$, $0.44$, $0.04$ for the three portfolios.

## 6.4 OPTIMAL PORTFOLIOS WITH POSITIVE WEIGHTS

Here we shall give a common formulation of the problem to determine the weights of the portfolios that have minimal variance or maximal drift subject to the constraint that the weights must not be negative. In Chapter 9 we shall use this formulation to give algorithms in the case the covariances have a simple form.

***Minimal Variance*** We shall minimize the variance $\boldsymbol{v} \cdot Q\boldsymbol{v}$ subject to the constraint $\boldsymbol{v} \in V$, where

$$V = \{\boldsymbol{v}; v_1 \geq 0, ..., v_m \geq 0, v_1 + \cdots + v_m = 1\}.$$

The set $V$ is compact and hence the minimum is attained in or on the boundary of $V$. That is, there is a set $I \subset \{1, 2, ..., m\}$ such that the minimum is attained at a point $\boldsymbol{v}$ for which $v_i > 0$ for $i \in I$ and $v_i = 0$ for $i \notin I$.

It follows from (6.8) and (6.9) that the variance is given by

$$1/\mathbf{1}_I \cdot Q_I^{-1}\mathbf{1}_I$$

and that the weights are proportional to

$$Q_I^{-1}\mathbf{1}_I.$$

Here $Q_I$ stands for the covariance matrix for the assets $i \in I$, and $\mathbf{1}_I$ has the corresponding dimension.

The set $I$ is thus the set that maximizes the quadratic form

$$\mathbf{1}_I \cdot Q_I^{-1}\mathbf{1}_I \text{ subject to the constraint } Q_I^{-1}\mathbf{1}_I \geq 0. \tag{6.51}$$

***Maximal Drift*** We shall here maximize the drift (6.16);

$$\nu_P = r_f + (\boldsymbol{\mu} - r_f\mathbf{1}) \cdot \boldsymbol{v} - \frac{1}{2}\boldsymbol{v} \cdot Q\boldsymbol{v} \tag{6.52}$$

subject to the constraint $\boldsymbol{v} \in V_1$, where

$$V_1 = \{\boldsymbol{v}; v_1 \geq 0, ..., v_m \geq 0\}.$$

This set is not compact, but (6.52) is dominated by

$$r_f + |\boldsymbol{\mu} - r_f\mathbf{1}||\boldsymbol{v}| - \frac{\lambda}{2}|\boldsymbol{v}|^2, \tag{6.53}$$

where $\lambda > 0$ is the smallest eigenvalue of $Q$.

Therefore $\nu_P < r_f$ for

$$|\boldsymbol{v}| > \frac{2|\boldsymbol{\mu} - r_f\mathbf{1}|}{\lambda}.$$

The maximum is thus attained on the compact set

$$V_1 \cap \{\boldsymbol{v}; |\boldsymbol{v}| \leq \frac{2|\boldsymbol{\mu} - r_f\mathbf{1}|}{\lambda}\}.$$

It follows from (6.27), (6.28), and (6.26) that the set $I$ is the set that maximizes the quadratic form

$$(\boldsymbol{\mu}_I - r_f\mathbf{1}_I) \cdot Q_I^{-1}(\boldsymbol{\mu}_I - r_f\mathbf{1}_I) \text{ subject to the constraint } Q_I^{-1}(\boldsymbol{\mu}_I - r_f\mathbf{1}_I) \geq 0. \tag{6.54}$$

***A Common Formulation*** Both (6.51) and (6.54) thus have the same form. Let $\boldsymbol{u} \in \mathbb{R}^m$, and let $\boldsymbol{u}_I = (u_i; i \in I)$. The problem is thus to determine the set $I$ that maximizes

$$\boldsymbol{u}_I \cdot Q_I^{-1}\boldsymbol{u}_I \text{ subject to the constraint } Q_I^{-1}\boldsymbol{u}_I \geq 0. \tag{6.55}$$

The minimum variance portfolio corresponds to $\boldsymbol{u} = \mathbf{1}$, and the maximal drift portfolio to $\boldsymbol{u} = \boldsymbol{\mu} - r_f\mathbf{1}$.

## 6.5 NOTES

The original works on portfolio selection are Markowitz (1952) and (1959)

# CHAPTER 7

# COVARIATION WITH THE MARKET

In order to construct the growth portfolios of the preceding chapter, we must be able to estimate the differences between the expected returns of the stocks and the short rate. The Capital Asset Pricing Model in this chapter reduces this problem to the problem of estimating the difference between the return of the market and the short rate. As a consequence, all portfolios are known if we can predict the return of the market, and the portfolio that has maximal drift for a given volatility is completely known if we can predict only the sign of this difference.

We also have a case study to illustrate how the portfolio theory given in this and the preceding chapter works in practice by formulating different scenarios and constructing the corresponding optimal portfolios.

*Mathematical Asset Management.* By Thomas Höglund

## 7.1 BETA

### 7.1.1 The Market

By the *market* or *market portfolio* in this context we shall mean an index. An index is a portfolio of assets. We shall enumerate these $1, ..., N$ in such a way that the first $m$ assets are those in our portfolio, and let $p_1, ...p_N$ denote the weights of the assets in the index.

Examples of Swedish stock indices are the OMX index, which consists of the 30 most traded stocks on the Stockholmsbörsen. Affärsvärldens Generalindex (AFGX) is a more extensive index that consists of about 300 stocks.

Examples of internationally well-known stock indices are Dow Jones (DJIA) and Standard and Poor 500 (S&P 500). There are also world indices that are very extensive portfolios.

### 7.1.2 Beta Value

The *beta value* of a stock, $\beta$, relative to a market is defined as

$$\beta = \mathrm{Cov}(R, R_M)/\sigma_M^2.$$

Here $R$ and $R_M$ are the returns of the stock and the market during a short period, a day for example, and $\sigma_M^2 = \mathrm{Var}(R_M)$. We shall use M = AFGX in our examples.

EXERCISE 7.1. Show that $\mathrm{Var}(R - bR_M)$ attains its minimum for $b = \beta$.

The drift during a period of length $\partial t$ equals $\nu \partial t$ whereas the volatility equals $\sigma\sqrt{\partial t}$. The drift during a short period therefore should be negligible compared to the volatility. That this is so is shown by Table 7.1, where data from Period 1– 4 in Table 4.1 have been used. The length of the period is one day, and the unit is %.

**Table 7.1.** Drift and volatility for *Five Stocks*

| | AFGX | AZN | LME | HM | SDIA | SKA |
|---|---|---|---|---|---|---|
| Drift | 0.09 | 0.07 | 0.18 | 0.14 | 0.22 | 0.07 |
| Volatility | 1.5 | 2.0 | 3.1 | 2.7 | 3.0 | 1.7 |

This fact and Exercise 7.1 explain the role of the beta value, and we thus have

$$R \approx \beta R_M. \tag{7.1}$$

Let

$$e = R - \beta R_M$$

and let $\rho$ denote the correlation coefficient between the returns of the stock and the market

$$\rho = \frac{\text{Cov}(R, R_M)}{\sigma \sigma_M},$$

where $\sigma$ is the volatility of the stock.

EXERCISE 7.2. Show that

$$\text{Var(R)} = \text{Var}(\beta R_M) + \text{Var(e)}$$

and that

$$\text{Var}(\beta R_M) = \rho^2 \sigma^2 \text{ and } \text{Var}(e) = (1 - \rho^2)\sigma^2.$$

The number $\rho^2$ is called the *coefficient of determination* and it is thus the proportion of the variance of $R$ that can be explained by the variation of the market.

Table 7.2 shows the beta values and coefficients of determination for *Five Stocks* during Period 1–4.

**Table 7.2.** Beta values and coefficient of determination for *Five Stocks*

| | AFGX | AZN | LME | HM | SDIA | SKA |
|---|---|---|---|---|---|---|
| $\beta$ | 1 | 0.43 | 1.46 | 0.73 | 1.07 | 0.42 |
| $\rho^2$ | 1 | 0.10 | 0.47 | 0.16 | 0.26 | 0.12 |

The *alpha value*, $\alpha$, of an asset is defined by

$$\alpha = Ee.$$

We thus have

$$R = \alpha + \beta R_M + e',$$

where $e'$ is uncorrelated with $R_M$ and has zero expectation. We have seen in Section 4.1.2 that it is hardly possible to estimate the return and hence $\alpha$ with sufficient precision for a single stock. It may, however, be possible to estimate the alpha for a fund.

EXERCISE 7.3. Let $\beta_i$ denote the beta value for asset $i$, $i = 1, ..., N$. Show that

$$p_1\beta_1 + \cdots + p_N\beta_N = 1.$$

## 7.2 PORTFOLIOS RELATED TO THE MARKET

### 7.2.1 The Beta Portfolio

Here we shall determine the weights for the portfolio that consists of given stocks and that maximizes the correlation with the market. Therefore let

$$\rho(\boldsymbol{v}) = \rho(R_P, R_M)$$

denote the correlation coefficient between the returns of the portfolio and the market. Here $\boldsymbol{v}$ stands for the weights.

EXERCISE 7.4. Show that

$$\rho(\boldsymbol{v}) = \frac{\boldsymbol{v} \cdot \boldsymbol{\beta}}{\sqrt{\boldsymbol{v} \cdot Q\boldsymbol{v}}} \sigma_M,$$

where $\boldsymbol{\beta} = (\beta_1, ..., \beta_m)$ are the beta values of the stocks in the portfolio.

We thus have

$$\rho(\boldsymbol{v}) = \frac{\boldsymbol{x} \cdot \boldsymbol{b}}{|\boldsymbol{x}|} \sigma_M,$$

where $\boldsymbol{x} = Q^{\frac{1}{2}}\boldsymbol{v}$ and $\boldsymbol{b} = Q^{-\frac{1}{2}}\boldsymbol{\beta}$.

The correlation is thus maximized for $\boldsymbol{x} = c\boldsymbol{b}$, that is, $\boldsymbol{v} = cQ^{-1}\boldsymbol{\beta}$, and $c > 0$. (The correlation is minimized if $c < 0$.)

The number $c$ is determined so that the weights sum to 1.

We shall call this portfolio the *Beta portfolio*. Observe that this portfolio exists only if $\mathbf{1} \cdot Q^{-1}\boldsymbol{\beta} > 0$.

EXERCISE 7.5. Assume that $\mathbf{1} \cdot Q^{-1}\boldsymbol{\beta} > 0$.

(a) Show that the Beta portfolio has the weights

$$\boldsymbol{v}_\beta = Q^{-1}\boldsymbol{\beta}/\mathbf{1} \cdot Q^{-1}\boldsymbol{\beta}, \tag{7.2}$$

the volatility

$$\sigma_\beta = \sqrt{\boldsymbol{\beta} \cdot Q^{-1}\boldsymbol{\beta}}/\mathbf{1} \cdot Q^{-1}\boldsymbol{\beta}, \tag{7.3}$$

and the beta value

$$\beta_\beta = \frac{\boldsymbol{\beta} \cdot Q^{-1}\boldsymbol{\beta}}{\mathbf{1} \cdot Q^{-1}\boldsymbol{\beta}}. \tag{7.4}$$

(b) Show that the correlation of the Beta portfolio with the market equals

$$\rho_{\beta,M} = \sigma_M \sqrt{\boldsymbol{\beta} \cdot Q^{-1}\boldsymbol{\beta}}$$

EXERCISE 7.6. (a) Show that the minimum variance portfolio has the beta value

$$\beta_{min} = \sigma_{min}^2 \mathbf{1} \cdot Q^{-1} \boldsymbol{\beta}. \tag{7.5}$$

(b) Show that the correlations of the minimum variance portfolio with the market and the Beta portfolio are

$$\rho_{\min,M} = \frac{\sigma_M}{\sigma_{min}} \beta_{min} \quad \text{and} \quad \rho_{\min,\beta} = \frac{\beta_{min}}{\sigma_{min}\sqrt{\boldsymbol{\beta} \cdot Q^{-1}\boldsymbol{\beta}}},$$

respectively.

It follows that

$$\rho_{\min,M} = \rho_{\min,\beta}\rho_{\beta,M}.$$

EXERCISE 7.7. Show that

$$\beta_{min} \leq \beta_\beta$$

with equality if and only if all beta values are equal.

Figure 7.1 shows the development of the Beta- and Minimum variance portfolios together with AFGX during Period 4. The portfolios are not rebalanced. In this case

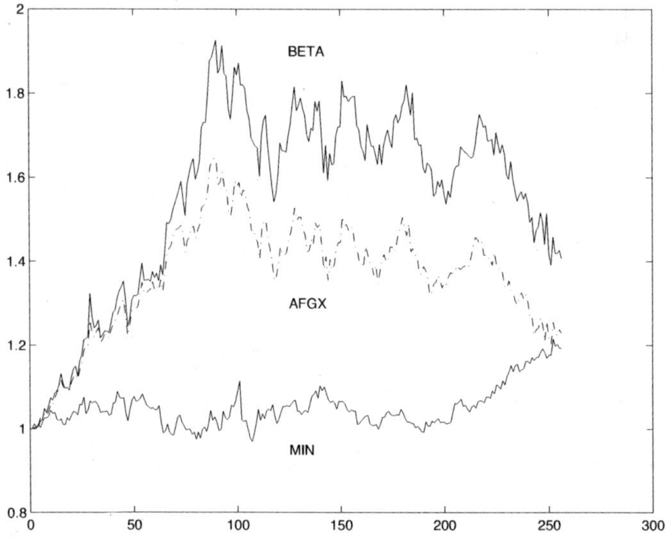


**Figure 7.1.** The development of the Beta- and Minimum variance portfolios during Period 4

the differences between the unchanged and the rebalanced portfolios are small. The weights were estimated with data from Period 1–3. The result is shown in Table 7.3.

**Table 7.3.** The weights of the Beta- and Minimum variance portfolios

| | AZN | LME | HM | SDIA | SKA | $\sigma$ | $\beta$ | $\rho^2$ | $\rho$ |
|---|---|---|---|---|---|---|---|---|---|
| $\beta$ | 0.63 | 1.49 | 0.75 | 1.11 | 0.58 | – | – | – | – |
| $\boldsymbol{v}_\beta$ | 0.20 | 0.40 | 0.11 | 0.16 | 0.13 | 0.29 | 1.05 | 0.58 | 0.76 |
| $\boldsymbol{v}_{min}$ | 0.35 | 0.01 | 0.17 | −0.01 | 0.47 | 0.23 | 0.64 | 0.35 | 0.59 |

The columns to the right show the volatilities, beta values, coefficients of determination, and correlations with the market during Period 1– 3.

The volatility of AFGX was $0.21$ during Period 1– 3. The unchanged Beta portfolio had the volatility $0.37$ during Period 4 (thus considerably higher than during Period 1– 3). The daily rebalanced Beta portfolio has the volatility $0.34$. The volatility of AFGX increased during Period 4 to $0.26$. Despite this increase of the market volatility, the volatilities of the Minimum variance portfolios did not increase, but remained $0.22$ whether or not the portfolios were rebalanced.

Figure 7.2 shows the development of the portfolios during Period 1– 4. The scale is logarithmic. Each unit on the vertical axis corresponds to a doubling/halving. The volatility of the Beta portfolio was $0.31$ to be compared to $0.23$ for AFGX and $0.22$ for the Minimum variance portfolio.

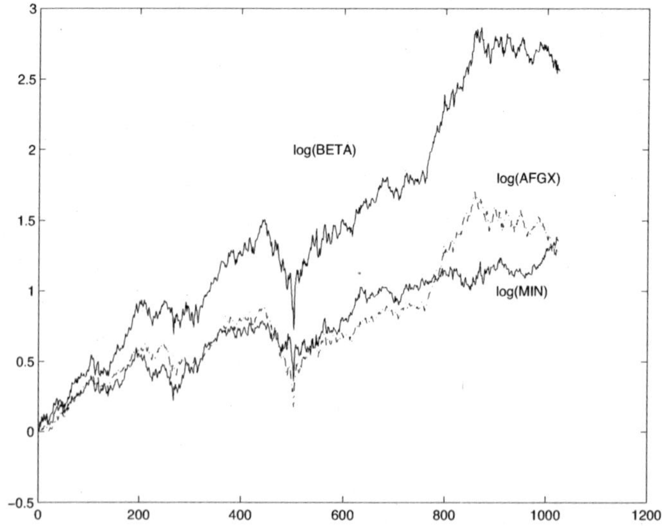


**Figure 7.2.** The development of the Beta- and Minimum variance portfolios during Period 1– 4. Logarithmic scale.

### 7.2.2 Stability of the Estimates of the Weights

The next theorem corresponds to Theorem 6.2.

**Theorem 7.1.** *Assume that the stock prices develop according to Model B in Section 4.3. Then the estimator $\hat{\boldsymbol{v}}_\beta$ of the weights (7.2) is, as $n \to \infty$, asymptotically normally distributed with expectation $\boldsymbol{v}_\beta$, and covariance matrix*

$$\frac{k}{n}\left(\sigma_{min}^2 Q^{-1} - \boldsymbol{v}_{min}\boldsymbol{v}_{min}^T + (\boldsymbol{v}_\beta - \boldsymbol{v}_{min})(\boldsymbol{v}_\beta - \boldsymbol{v}_{min})^T\right). \tag{7.6}$$

*Here*

$$k = \frac{1 - \sigma_M^2 \boldsymbol{\beta}\cdot Q^{-1}\boldsymbol{\beta}}{\sigma_{min}^2 \sigma_M^2 (\mathbf{1}\cdot Q^{-1}\boldsymbol{\beta})^2}.$$

*Proof.* Let $\boldsymbol{q} = (q_1, ..., q_m)^T$, where $q_i = \sigma_{i,M} = \text{Cov}(R_i, R_M)$. Then the weights (7.2) satisfy

$$\boldsymbol{v}_\beta = Q^{-1}\boldsymbol{q}/\mathbf{1}\cdot Q^{-1}\boldsymbol{q}.$$

If we put $\boldsymbol{z}_n = \sqrt{n}(\hat{\boldsymbol{q}} - \boldsymbol{q})$ and proceed as in the proof of Theorem 6.2 we see that $\hat{\boldsymbol{v}}_\beta$ has the same asymptotic distribution as

$$\boldsymbol{v}_\beta - \frac{1}{\sqrt{n}}\boldsymbol{u}_n.$$

Here $\boldsymbol{u}_n \to \boldsymbol{u}$, and $\boldsymbol{u}$ is as in (6.12), but with $\boldsymbol{v} = \boldsymbol{v}_\beta$, and

$$\boldsymbol{w} = \boldsymbol{Z}\boldsymbol{v} - \boldsymbol{z}/\mathbf{1}\cdot Q^{-1}\boldsymbol{q}.$$

In this case

$$\boldsymbol{u}\boldsymbol{u}^T = Q^{-1}\boldsymbol{w}\boldsymbol{w}^T Q^{-1} - Q^{-1}\boldsymbol{w}\boldsymbol{w}^T Q^{-1}\mathbf{1}\boldsymbol{v}^T - \boldsymbol{v}(Q^{-1}\mathbf{1})^T\boldsymbol{w}\boldsymbol{w}^T Q^{-1} + \boldsymbol{v}(Q^{-1}\mathbf{1})^T\boldsymbol{w}\boldsymbol{w}^T Q^{-1}\mathbf{1}\boldsymbol{v}^T. \tag{7.7}$$

We have

$$\boldsymbol{w}\boldsymbol{w}^T = \boldsymbol{Z}\boldsymbol{v}\boldsymbol{v}^T\boldsymbol{Z} - \boldsymbol{z}\boldsymbol{v}^T\boldsymbol{Z}/\mathbf{1}\cdot Q^{-1}\boldsymbol{q} - \boldsymbol{Z}\boldsymbol{v}\boldsymbol{z}^T/\mathbf{1}\cdot Q^{-1}\boldsymbol{q} + \boldsymbol{z}\boldsymbol{z}^T/(\mathbf{1}\cdot Q^{-1}\boldsymbol{q})^2$$

and

$$E[\boldsymbol{z}_i(\boldsymbol{Z}\boldsymbol{v})_j] = \sum_l E[z_i Z_{j,l}]v_l = \sum_l (\sigma_{i,j}\sigma_{M,l} + \sigma_{i,l}\sigma_{M,j})v_l = \sigma_{i,j}\boldsymbol{q}\cdot\boldsymbol{v} + (Q\boldsymbol{v})_i q_j.$$

In a similar way we get

$$E[(\boldsymbol{Z}\boldsymbol{v})_i\boldsymbol{z}_j] = \sigma_{i,j}\boldsymbol{q}\cdot\boldsymbol{v} + q_i(Q\boldsymbol{v})_j,$$

and

$$E[z_i z_j] = \sigma_{i,j}\sigma_M^2 + q_i q_j.$$

It follows from this and (6.15) that $E[\boldsymbol{w}\boldsymbol{w}^T] = AQ$, where

$$A = \frac{\sigma_M^2 - \boldsymbol{q} \cdot Q^{-1}\boldsymbol{q}}{(\boldsymbol{1} \cdot Q^{-1}\boldsymbol{q})^2}.$$

The theorem now follows from (7.7) and the identity $k = A/\sigma^2_{min}$. □

The weights of the Beta portfolio for *Five Stocks* estimated during the different periods are given in Table 7.4.

**Table 7.4.** The weights of the Beta portfolio during different periods

| Period | AZN | LME | HM | SDIA | SKA |
|---|---|---|---|---|---|
| 1 | 0.23 | 0.44 | 0.03 | 0.14 | 0.16 |
| 2 | 0.13 | 0.40 | 0.13 | 0.18 | 0.16 |
| 3 | 0.22 | 0.34 | 0.19 | 0.17 | 0.08 |
| 4 | 0.09 | 0.49 | 0.19 | 0.19 | 0.04 |
| 1– 2 | 0.17 | 0.42 | 0.08 | 0.17 | 0.16 |
| 3– 4 | 0.14 | 0.43 | 0.18 | 0.18 | 0.07 |
| 1– 4 | 0.15 | 0.44 | 0.15 | 0.18 | 0.09 |

Define $d_{\text{theor}}$ and $d_{\text{obs}}$ in the same way as for the Minimum variance portfolio. These distances are given in Tables 7.5 and 7.6 for a number of different period lengths. The estimates are based on the daily growths in Table 7.5 and on the weekly growths in Table 7.6.

**Table 7.5.** Distance between the weights. Daily growths

| Period Length | Number of Periods | $\hat{d}_{\text{theor}}$ | $d_{\text{obs}}$ | $d_{\text{obs}}/\hat{d}_{\text{theor}}$ |
|---|---|---|---|---|
| 1024 | 1 | 0.03 | – | – |
| 512 | 2 | 0.04 | 0.05 | 1.20 |
| 256 | 4 | 0.05 | 0.05 | 0.99 |
| 128 | 8 | 0.22 | 0.15 | 0.69 |
| 64 | 16 | 213.42 | 5.99 | 0.03 |

The reason for the large deviation at the period length 64 is that the estimate of $\beta_{min}$ became very small during one of the periods.

In conformity with the Minimum variance portfolio, the weights of this portfolio seem to be reasonably stable if the estimates are based on data from some years.

**Table 7.6.** Distance between the weights. Weekly growths

| Period Length | Number of Periods | $\hat{d}_{\text{theor}}$ | $d_{\text{obs}}$ | $d_{\text{obs}}/\hat{d}_{\text{theor}}$ |
|---|---|---|---|---|
| 1020 | 1 | 0.04 | – | – |
| 510 | 2 | 0.06 | 0.06 | 0.96 |
| 255 | 4 | 0.08 | 0.09 | 1.12 |
| 125 | 8 | 0.15 | 0.17 | 1.14 |
| 65 | 15 | 0.27 | 0.20 | 0.73 |

### 7.2.3 Market Neutral Portfolios

This idea goes back to Alfred Jones, who started a hedge found in 1949. The new idea was that he not only bought stocks he considered undervalued but also borrowed and sold stocks he considered overvalued.

When the stock market rose, so did the value of the long positions, while the value of the short positions fell. The opposite occurred when the stock market fell.

In this way the portfolio became less dependent on market fluctuations and more dependent on Jones' own skill in analyzing individual stocks.

We shall say that a portfolio is *market neutral* if the return of the portfolio is uncorrelated with the return of the market;

$$\text{Cov}(R_P, R_M) = 0, \quad \text{i.e.,} \quad v_1\beta_1 + \cdots + v_m\beta_m = 0.$$

EXERCISE 7.8. Show that a portfolio is market neutral if and only if it is uncorrelated with the Beta portfolio.

EXERCISE 7.9. Show that the market neutral portfolio that has minimum variance has the weights

$$(1+\gamma)\boldsymbol{v}_{min} - \gamma\boldsymbol{v}_{\beta},$$

where

$$\gamma = \frac{\sigma_{min}^2}{\sigma_\beta^2 - \sigma_{min}^2}.$$

Figure 7.3 shows the development of this portfolio during Period 4.

The weights were estimated with data from Period 1– 3. The calculated volatility is 0.36. In this case we got $\gamma = 1.54$, and the portfolio has the weights 0.60, −0.58, 0.27, −0.27, 0.98. Note the great difference between the unchanged and the daily or continuously rebalanced portfolio in this case.

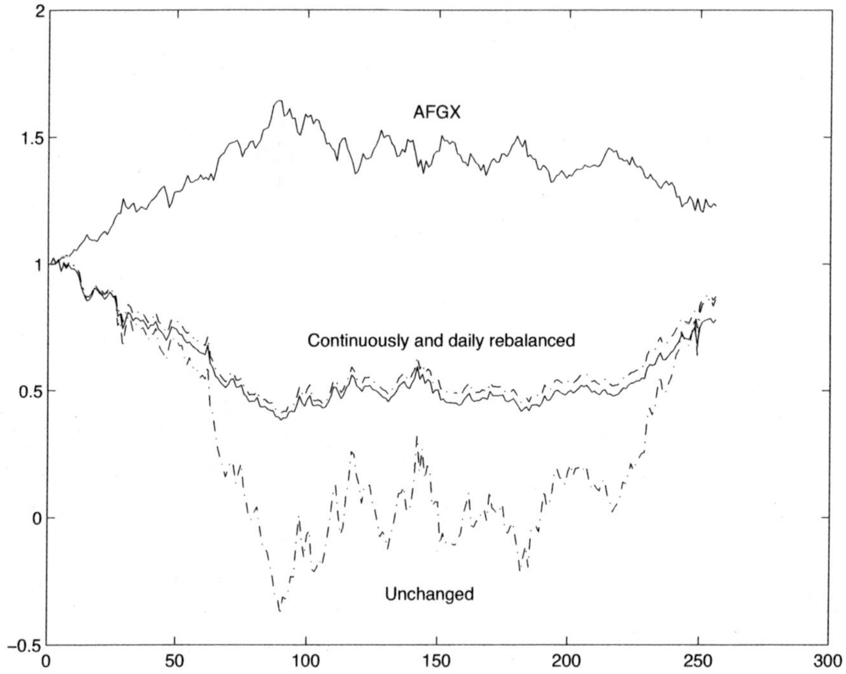


**Figure 7.3.** Development during Period 4 of AFGX and the market neutral portfolio that has minimum variance.

## 7.3 CAPITAL ASSET PRICING MODEL

### 7.3.1 The CAPM Identity

To begin with, consider portfolios that consist of all $N$ assets on the market. According to an economic theory called the Capital Asset Pricing Model (CAPM), the market portfolio and the Markowitz portfolio (6.31) coincide. Recall that the Markowitz portfolio is the stock portfolio in the portfolio that has maximal expected growth.

Thus

$$\frac{\sigma_{min}^2 Q^{-1}(\boldsymbol{\mu} - r_f \mathbf{1})}{\mu_{min} - r_f} = \boldsymbol{p}. \tag{7.8}$$

Here $\boldsymbol{p} = (p_1, ..., p_N)$ are the weights of the market portfolio, and $Q$ is a $N \times N$ matrix. It follows that

$$\boldsymbol{\mu} - r_f \mathbf{1} = \frac{\mu_{min} - r_f}{\sigma_{min}^2} Q\boldsymbol{p}. \tag{7.9}$$

We have

$$\text{Cov}(R_i, R_M) = \sum_{j=1}^{N} \sigma_{i,j} p_j = (Q\boldsymbol{p})_i \tag{7.10}$$

and hence

$$\sigma_M^2 \boldsymbol{\beta} = Q\boldsymbol{p}. \tag{7.11}$$

It follows from (7.9) and (7.11) that

$$\boldsymbol{\mu} - r_f \mathbf{1} = \Delta \boldsymbol{\beta}, \tag{7.12}$$

where

$$\Delta = \frac{(\mu_{min} - r_f)\sigma_M}{\sigma_{min}^2}. \tag{7.13}$$

It follows from Exercise 7.3 that $\boldsymbol{p} \cdot \boldsymbol{\beta} = 1$, and hence by (7.12)

$$\Delta = \mu_M - r_f, \tag{7.14}$$

where $\mu_M = \boldsymbol{p} \cdot \boldsymbol{\mu}$ is the expected return of the market.

We have thus shown that CAPM implies the *CAPM identity*

$$\mu_i - r_f = \beta_i(\mu_M - r_f) \quad \text{for } i = 1, ..., N. \tag{7.15}$$

The beta values are defined by covariances that can be estimated with historical data, and hence the beta values can be considered as known. The CAPM identity thus reduces the problem of estimating the expected return of individual stocks to the problem of estimating the expected return of the market.

Also, as we shall see below, the CAPM identity implies that the weights of the Markowitz portfolio and the portfolio (6.34) that has maximal drift for a given portfolio volatility will not depend on $\Delta$ but just $\boldsymbol{\beta}$ and $Q$, provided that $\Delta > 0$. These portfolios are therefore known if the CAPM identity holds, even if we cannot predict the expected return of the market with higher precision than $\Delta > 0$.

It is therefore of interest to investigate how well the CAPM identity agrees with reality.

The CAPM identity is a statement concerning expected returns,

$$E(R_i - \beta_i R_M) = r_f(1 - \beta_i),$$

and we know from Chapter 4 that we cannot make any precise statements about this expected value because the volatility is too big. The role of beta is, however, explained by (7.1).

Here we shall examine whether there is a positive connection between returns and beta values for *Five Stocks*. Data from the four periods are shown in Table 7.7.

**Table 7.7.** Returns and Beta values for *Five Stocks*

| Period | | AFGX | AZN | LME | HM | SDIA | SKA |
|---|---|---|---|---|---|---|---|
| 1 | Return | 0.39 | 0.18 | 0.79 | 0.68 | 0.72 | 0.19 |
| | Beta | 1 | 0.96 | 1.36 | 0.60 | 0.98 | 0.65 |
| 2 | Return | −0.04 | 0.02 | 0.03 | 0.61 | 0.45 | 0.14 |
| | Beta | 1 | 0.44 | 1.20 | 0.72 | 1.19 | 0.53 |
| 3 | Return | 0.39 | 0.46 | 0.82 | 0.52 | 0.67 | 0.23 |
| | Beta | 1 | 0.83 | 2.09 | 0.92 | 1.23 | 0.64 |
| 4 | Return | 0.24 | 0.29 | 0.56 | 0.00 | 0.80 | 0.27 |
| | Beta | 1 | 0.09 | 1.45 | 0.70 | 1.02 | 1.14 |
| 1– 4 | Return | 0.24 | 0.23 | 0.56 | 0.45 | 0.67 | 0.20 |
| | Beta | 1 | 0.43 | 1.46 | 0.73 | 1.07 | 0.42 |

Note that the return of AFGX during Period 2 is lower than the interest rate, and hence the stock with the highest beta value has the lowest expected return according to CAPM.

In Table 7.8 we have ranked the returns and the beta values of the stocks in order to get an idea of whether there is a connection.

**Table 7.8.** Ranked returns and Beta values for *Five Stocks*

| Period | | AZN | LME | HM | SDIA | SKA | $\rho$ |
|---|---|---|---|---|---|---|---|
| 1 | Return | 5 | 1 | 3 | 2 | 4 | |
| | Beta | 3 | 1 | 5 | 2 | 4 | 0.6 |
| 2 | Return | 5 | 4 | 1 | 2 | 3 | |
| | beta | 5 | 1 | 3 | 2 | 4 | 0.3 |
| 3 | Return | 4 | 1 | 3 | 2 | 5 | |
| | Beta | 4 | 1 | 3 | 2 | 5 | 1 |
| 4 | Return | 3 | 2 | 5 | 1 | 4 | |
| | Beta | 5 | 1 | 4 | 3 | 2 | 0.3 |
| 1– 4 | Return | 4 | 2 | 3 | 1 | 5 | |
| | Beta | 4 | 1 | 3 | 2 | 5 | 0.9 |

The rightmost column shows Spearman's rank correlation coefficient, $\rho$: Let $x_1, ..., x_m$ and $y_1, ..., y_m$ denote the ranks of the $m = 5$ stocks during a period. Then

$$\rho = 1 - \frac{6}{m(m^2-1)} \sum_{i=1}^{m} (x_i - y_i)^2.$$

Let $\rho_1, ..., \rho_n$ denote the rank correlation coefficients during the $n = 4$ periods, and let $\check{\rho}_i = \rho_i$ if the return of the market exceeded the interest rate during period $i$, and let $\check{\rho}_i = -\rho_i$ otherwise. If (7.1) holds then $\check{\rho}_1, ..., \check{\rho}_n$ tend to be positive. On the other hand, if there is no connection but the ranks are random permutations of the numbers $1, ..., m$, then

$$(\check{\rho}_1 + \cdots + \check{\rho}_n)\sqrt{\frac{m-1}{n}} \tag{7.16}$$

is approximately normally distributed with zero expectation and variance 1. In our example, (7.16) has the value

$$2.5\sqrt{\frac{5-1}{4}} = 2.5.$$

We have $1 - \Phi(2.5) = 0.006$, and hence there seems to be a positive connection between returns and beta values.

**Summary 7.2.** *CAPM implies the CAPM identity (7.15). We cannot verify or falsify this identity, but there is some evidence for (7.1).*

### 7.3.2 Consequences of CAPM

Here we shall assume that the CAPM identity (7.15) holds, and we will see which consequences this yields.

A first consequence is the following:

EXERCISE 7.10. Show that the expected return of a portfolio with weights $\boldsymbol{v}$ equals

$$r_f + \beta_{\boldsymbol{v}}\Delta. \tag{7.17}$$

Here $\beta_{\boldsymbol{v}} = \boldsymbol{\beta} \cdot \boldsymbol{v}$ is the beta value of the portfolio, and $\Delta = \mu_M - r_f$.

Another consequence is that the Markowitz portfolio, (6.31), and the Beta portfolio coincide.

EXERCISE 7.11. Assume that $\Delta > 0$ and $\beta_{min} > 0$. Show that $\boldsymbol{v}_{Mz} = \boldsymbol{v}_{\beta}$.

We next turn to the portfolios (6.26), (6.34), and (6.38).

EXERCISE 7.12. Show that

(a)

$$\boldsymbol{v}_{max} = \Delta Q^{-1}\boldsymbol{\beta},\ \sigma_{max}^2 = \Delta^2\boldsymbol{\beta} \cdot Q^{-1}\boldsymbol{\beta},\ \omega = \Delta\mathbf{1} \cdot Q^{-1}\boldsymbol{\beta}, \tag{7.18}$$

(b)

$$\frac{\sigma}{\sigma_{max}}\boldsymbol{v}_{max} = \frac{\text{sign}(\Delta)\sigma}{\sqrt{\boldsymbol{\beta} \cdot Q^{-1}\boldsymbol{\beta}}}Q^{-1}\boldsymbol{\beta}, \tag{7.19}$$

(c)

$$\boldsymbol{v}_{aux} = \Delta\frac{\beta_{min}}{\sigma_{min}^2}(\boldsymbol{v}_{\beta} - \boldsymbol{v}_{min}),\ \sigma_{aux}^2 = \Delta^2(\boldsymbol{\beta} - \beta_{min}\mathbf{1}) \cdot Q^{-1}(\boldsymbol{\beta} - \beta_{min}\mathbf{1}). \tag{7.20}$$

Note that the portfolio (7.19) does not depend on $\Delta$ as long as $\Delta > 0$. This portfolio is therefore known even if we cannot estimate the expected return of the market with higher precision than that $\Delta > 0$ .

The maximal growth portfolio thus has the weight $\omega = \Delta\beta_{min}/\sigma^2_{min}$ in the Beta portfolio and the remainder in cash, provided that $\Delta \leq h\sigma^2_{min}/\beta_{min}$, where $h$ is the maximal total stock weight. Otherwise the stock portfolio has the weights

$$\boldsymbol{v}_{max}(h,\Delta) = h\boldsymbol{v}_{min} + \boldsymbol{v}_{aux}.$$

For use below we have marked here the dependence on both $h$ and $\Delta$.

The weights thus depend on (a) what we think about the future development on the stock market and (b) if we are allowed to pledge the portfolio or not, that is, on $\Delta$ and $h$.

Below we shall illustrate how the portfolio theory given in this and the preceding chapter works in practice by formulating different scenarios and constructing the corresponding optimal portfolios.

***A Case Study*** Assume that we are at the end of Period 3 and consider three scenarios:

Case A. We believe in a moderate development on the stock market, $\Delta = 0.05$.

Case B. We believe in a good development, $\Delta = 0.10$, but do not want to pledge the stock portfolio.

Case C. We believe in a strong development, $\Delta = 0.35$, and accept to use maximal pledge.

We based the estimates on data from *Five Stocks* during Period 1– 3 and got

$$\frac{\beta_{min}}{\sigma^2_{min}}(\boldsymbol{v}_\beta - \boldsymbol{v}_{min}) = (-1.97, 4.77, -0.77, 2.13, -4.16).$$

In case A we have $\omega = 0.62$, and hence $62\%$ will be invested in the Beta portfolio and the remainder in cash. The development of this portfolio during Period 4 is shown in Fig. 7.4. We have assumed that the interest rate equals $4\%$. The volatility of the portfolio was $0.21$ and the return was $32\%$, to be compared to $0.28$ and $23\%$ for AFGX, respectively.

In case B we have $\omega = 1.24$. Since $\omega > 1$ and the portfolio must not be pledged, we shall place all the money in the stock portfolio that has the weights

$$\boldsymbol{v}_{\text{max}}(1, 0.10) = (0.16, 0.49, 0.10, 0.21, 0.05).$$

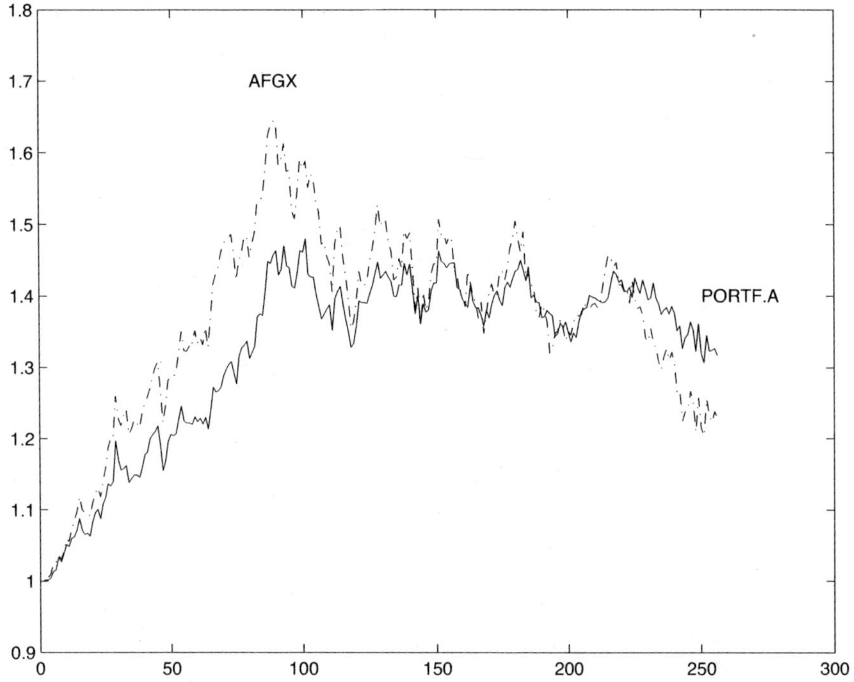


**Figure 7.4.** The development of Portfolio A during Period 4.

The volatility was $0.39$ and the return was $54\%$. See Fig. 7.5.

In case C we have $\omega = 4.35$, and the maximal total stock weight $3.15 = 315\%$. We will thus have $-2.15$ in cash and $3.15$ in the stock portfolio, which has the (relative) weights

$$\boldsymbol{v}_{\max}(3.15, 0.35)/3.15 = (0.13, 0.54, 0.09, 0.23, 0.00).$$

See Fig. 7.6. We have assumed that we pay $7\%$ interest on the money we borrow.

The volatility became $1.32 = 132\%$, and the return was $90\%$. It is seen that the pledge gives leverage to the portfolio development. Both up and down.

It is necessary to rebalance a maximally pledged portfolio if the portfolio value falls. All three portfolios have been rebalanced daily.

In Figs. 7.7, 7.8, and 7.9, the development of these portfolios during the entire Period 1– 4 are shown.

It is hard to read the portfolio development in the beginning in Fig. 7.9. A plot with logarithmic scale on the vertical axis becomes more clear. See Fig. 7.10.

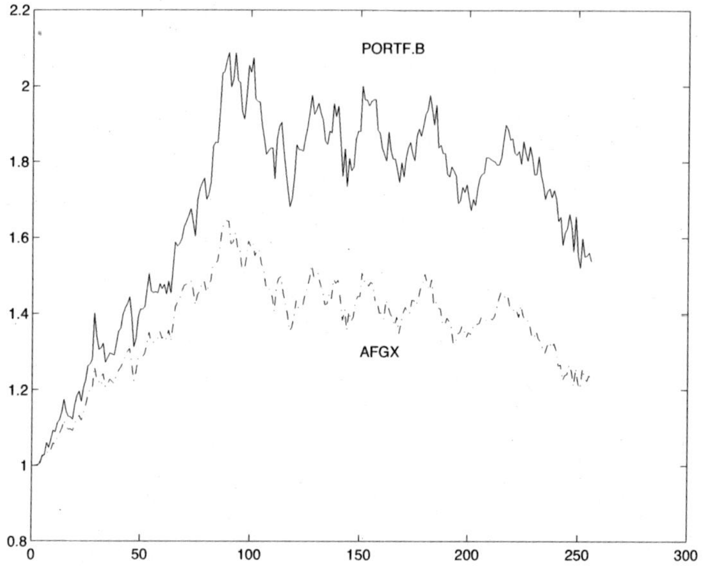


**Figure 7.5.** The development of Portfolio B during Period 4.

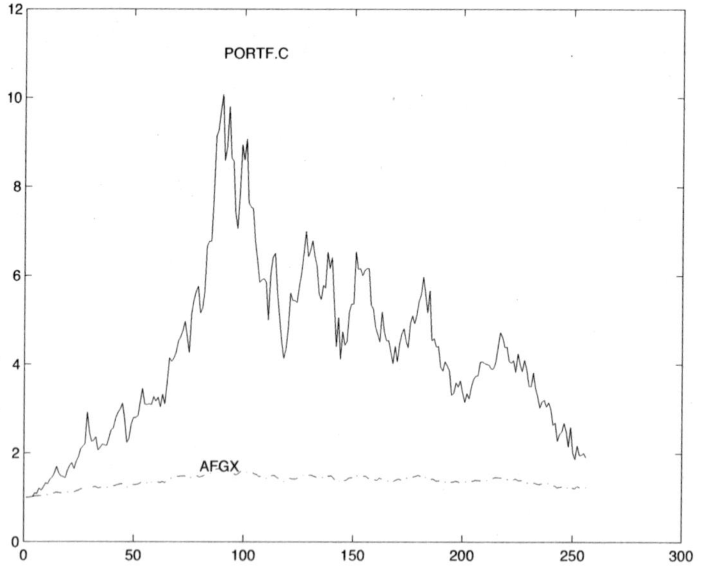


**Figure 7.6.** The development of Portfolio C during Period 4.

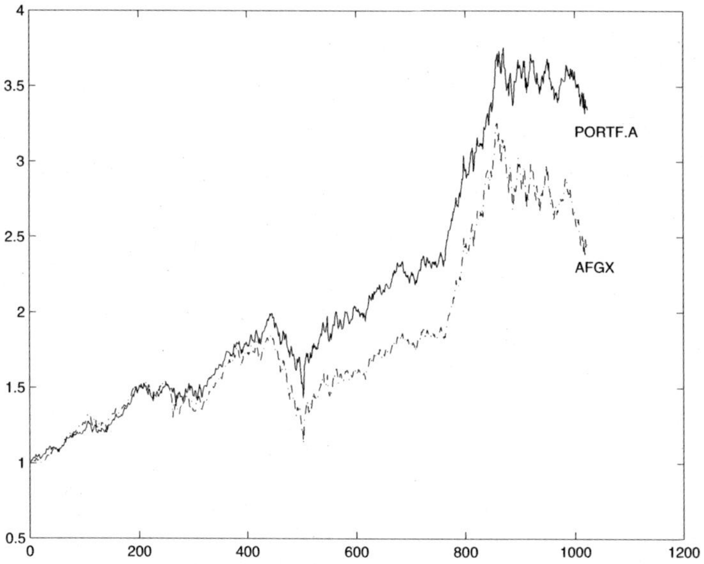


**Figure 7.7.** The development of Portfolio A during Period 1– 4.

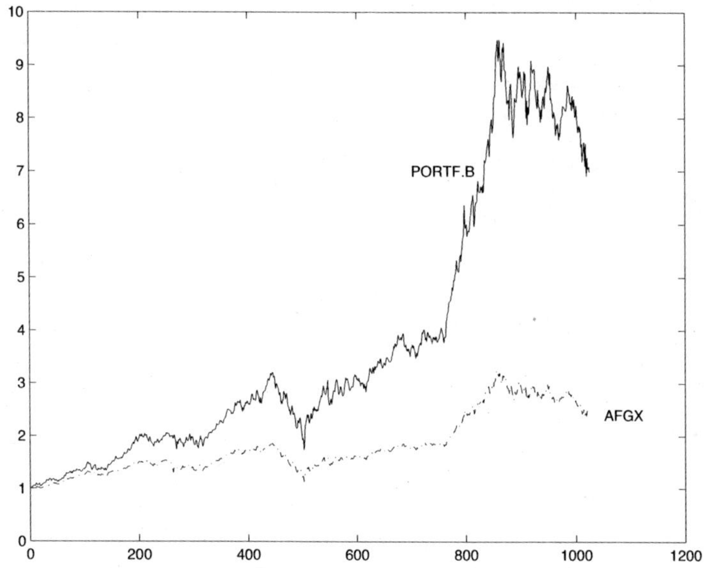


**Figure 7.8.** The development of Portfolio B during Period 1– 4.

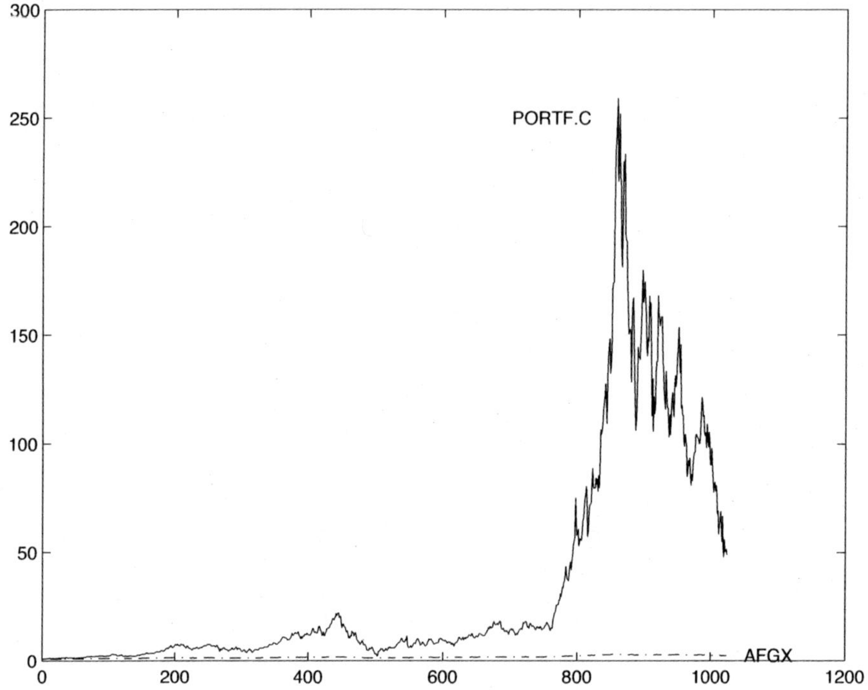


**Figure 7.9.** The development of Portfolio C during Period 1– 4.

Table 7.9 shows the volatilities and yearly growth factors for the portfolios. (The yearly growth factor in the interval $(0, T)$ is

$$(\frac{P(T)}{P(0)})^{\frac{1}{T}},$$

where $P(0)$ and $P(T)$ are the values of the portfolio at time 0 and $T$ years, respectively.)

**Table 7.9.** Volatilities and growth factors for portfolios A, B and C

| | AFGX | PORTF.A | PORTF.B | PORTF.C |
|---|---|---|---|---|
| Volatility | 0.23 | 0.19 | 0.35 | 1.17 |
| Growth factor | 1.24 | 1.34 | 1.61 | 2.59 |

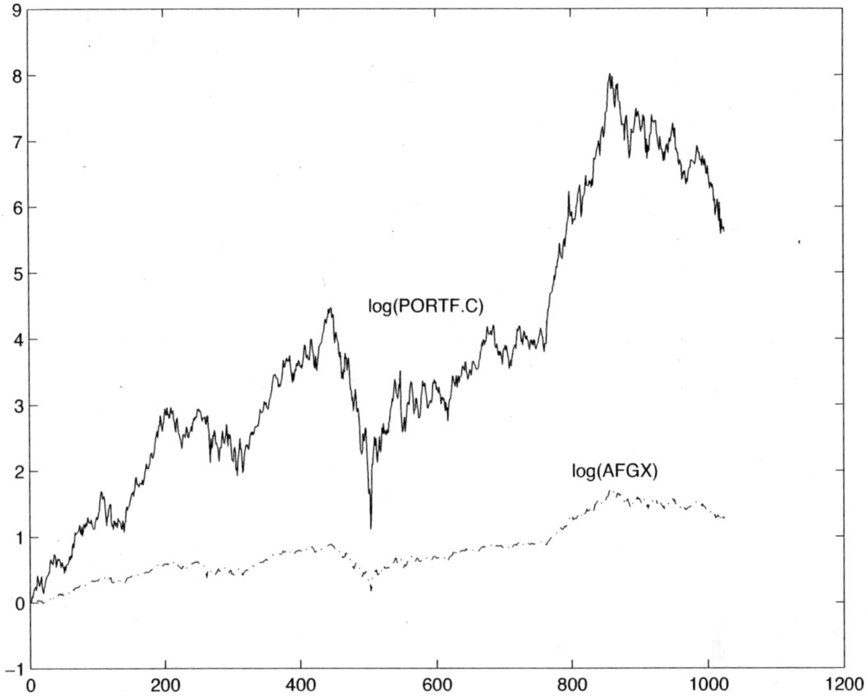


**Figure 7.10.** The development of Portfolio C during Period 1– 4. Logarithmic scale.

### 7.3.3 The Market Portfolio

We shall here consider all $N$ stocks in the index. The vector with the weights of the stocks in the index is denoted $\boldsymbol{p} = (p_1, ..., p_N)$. The matrix $Q$ has the dimension $N \times N$, the vector $\boldsymbol{\beta}$ has the dimension $N$, and so on. The CAPM identity is assumed to hold.

EXERCISE 7.13. According to CAPM, the market portfolio and the Markowitz portfolio coincide. That is,

$$\frac{Q^{-1}\boldsymbol{\beta}}{\mathbf{1} \cdot Q^{-1}\boldsymbol{\beta}} = \boldsymbol{p}.$$

Verify this directly; that is, show that this is a mathematical identity.

EXERCISE 7.14. Show that

$$\beta_{min} = \frac{\sigma^2_{min}}{\sigma^2_M}, \quad \omega = \frac{\Delta}{\sigma^2_M} \text{ and } \sigma^2_{max} = \frac{\Delta^2}{\sigma^2_M}.$$

Maximal growth is obtained by placing the weight $\omega$ in the market portfolio and placing the remainder in cash, provided that $\omega \leq h$. According to Exercise 7.14b, this does not hold if $\Delta > h\sigma_M^2$.

If $\sigma_M = 0.23$ and if $h = 1$ — that is, we are not allowed to pledge the stock portfolio — this inequality takes the form $\Delta > 0.05$. If we are allowed to use maximal pledge and if $1/(1 - \mathbf{c} \cdot \mathbf{p}) \approx 3$, then the inequality instead becomes $\Delta > 0.16$.

We have thus made the following observation.

**Observation 7.3.** *Assume that the CAPM identity holds. The portfolio that gives maximal growth has part of the money in cash and has the remainder in the market portfolio, provided that $\Delta \leq h\sigma_M^2$. Otherwise the stock portfolio and the market portfolio do not coincide. The latter holds in particular in the following two cases.*

*(a) If we are not allowed to pledge the stock portfolio and we believe in a decent development on the stock market, $\Delta > \sigma_M^2$.*

*(b) If we are allowed to use maximal pledge and we believe in a strong development on the stock market, $\Delta > \sigma_M^2/(1 - \mathbf{c} \cdot \mathbf{p})$.*

## 7.4 NOTES

To what extent CAPM describes the real world is discussed in Perold (2004) and in Fama and French (2004).

# CHAPTER 8

# PERFORMANCE AND RISK MEASURES

## 8.1 PERFORMANCE MEASURES

Let $S_A(t)$ and $S_B(t)$ denote the values of two assets (or portfolios) in $t$ years. We shall try to decide which of these that will perform best. The time $t$ must be large if we shall be able to do this with some certainty. We have seen in Chapter 4 that the growths, $G_A(0,t)$ and $G_B(0,t)$, in this case often are approximately normally distributed, and we shall in this section assume that Model B holds.

The probability that $A$ will have performed better than $B$ at time $t$ then equals

$$P(\frac{S_A(t)}{S_A(0)} > \frac{S_B(t)}{S_B(0)}) = \Phi(P_r\sqrt{t}), \tag{8.1}$$

where

$$P_r = \frac{E[G_A(0,1) - G_B(0,1)]}{\sqrt{\text{Var}(G_A(0,1) - G_B(0,1))}}. \tag{8.2}$$

This is so because

$$S_A(t)/S_A(0) > S_B(t)/S_B(0)$$

if and only if

$$G_A(0,t) - G_B(0,t) > 0.$$

The ratio $P_r$ can also be written as in the next exercise.

EXERCISE 8.1. Let $\nu_i$ and $\sigma_i$ denote the drift and the volatility for the asset $i$, $i = A, B$. In addition; let $\rho$ denote the correlation coefficient between $G_A(0,1)$ and $G_B(0,1)$. Show that

$$P_r = \frac{\nu_A - \nu_B}{\sqrt{\sigma_A^2 + \sigma_B^2 - 2\rho\sigma_A\sigma_B}}.$$

We shall call $P_r$ for the *performance ratio* for $A$ relative to $B$. The probability that $A$ performs better than $B$ is thus an increasing function of $P_r$. If $P_r > 0$, then this probability is $> 1/2$, and it tends to 1 as $t \to \infty$.

Given a probability $\epsilon$, we can give a time $T_\epsilon$ such that the probability that A has performed better than B at time $T_\epsilon$ equals $1 - \epsilon$. Namely, $T_\epsilon = (z_\epsilon/P_r)^2$.

The performance ratio is positive if and only if $\nu_A > \nu_B$. It is therefore clear that

$$\text{if} \quad P_r(A,B) > 0 \quad \text{and} \quad P_r(B,C) > 0, \quad \text{then} \quad P_r(A,C) > 0.$$

On the other hand, it is possible that

$$P_r(B,C) > P_r(A,C) \quad \text{but} \quad P_r(A,B) > 0.$$

***Example 8.1.*** To illustrate this, let $A$ be a stock portfolio with $\nu_A = 0.08$ and $\sigma_A = 0.20$, and let $B$ be a bond portfolio with $\nu_B = 0.04$ and $\sigma_B = 0.03$, whereas $C$ is money in a desk drawer, $\nu_C = 0$ and $\sigma_C = 0$. If we assume that the stock and bond portfolios are uncorrelated, then we get

$$P_r(A,B) = 0.20, \quad P_r(B,C) = 1.33, \quad \text{and} \quad P_r(A,C) = 0.40.$$

The probability that the stock portfolio has performed better than the bond portfolio after 4 years is therefore about 65%. Thus this does not prevent the fact that the probability that the value of the stock portfolio decreases, $\approx 21\%$, is essentially larger than the probability that the bond portfolio decreases, $\approx 0.5\%$. □

An estimate of $P_r$ based on observations at $0, \partial t, ..., n\partial t$, where $n\partial t = T$, is given by

$$\hat{P}_r = \frac{(G_A(0,T) - G_B(0,T))/T}{\hat{\sigma}}. \tag{8.3}$$

Here

$$\hat{\sigma}^2 = \sum_{i=1}^{n}(x_i - \bar{x})^2/T, \text{ where } x_i = G_A((i-1)\partial t, i\partial t) - G_B((i-1)\partial t, i\partial t).$$

EXERCISE 8.2. Show that $\hat{P}_r$ is asymptotically normally distributed, as $n \to \infty$, with expectation $P_r$ and variance

$$\frac{1}{T}\Big(1 + \frac{P_r^2}{2}\partial t\Big).$$

*Hint.* Use Exercise 4.5 and the fact that the numerator and denominator in (8.3) are stochastically independent.

A confidence interval for $P_r$ with the approximative confidence level $1 - \epsilon$ is thus given by

$$\hat{P}_r \pm z_{\epsilon/2}\sqrt{\frac{1}{T}(1 + \frac{\hat{P}_r^2}{2}\partial t)} \approx \hat{P}_r \pm z_{\epsilon/2}\sqrt{\frac{1}{T}}, \tag{8.4}$$

where the interval to the right can be used when $\hat{P}_r$ and $\partial t$ are such that $\frac{\hat{P}_r^2}{2}\partial t$ is negligible compared to 1.

If, in the first place, we are interested in the sign of $P_r$ it can be appropriate to look at an one-sided confidence interval with the lower bound

$$\hat{P}_r - z_\epsilon\sqrt{\frac{1}{T}}.$$

This interval is a subset of the interval $(0, \infty)$ if and only if the confidence level $1 - \epsilon$ is less than

$$\Phi(\sqrt{T}\hat{P}_r). \tag{8.5}$$

This value is thus the upper bound for the confidence levels of the intervals that are subsets of $(0, \infty)$. It is also an estimate of the probability (8.1) with $t = T$.

Note that $A$ and $B$ can be two portfolios consisting of a number of assets with given weights, $(\gamma_A, \boldsymbol{v}_A)$ and $(\gamma_B, \boldsymbol{v}_B)$, respectively. Here $\gamma_A$ and $\gamma_B$ are the weights of the cash holdings, and $\boldsymbol{v}_A$ and $\boldsymbol{v}_B$ the stock weights. The growths of the portfolios are given by (6.49), provided that the portfolios are rebalanced in such a way that the weights are restored:

$$G(0,t) = (\gamma r_f + \boldsymbol{v}\cdot\mu - \frac{1}{2}\boldsymbol{v}\cdot Q\boldsymbol{v})t + \boldsymbol{v}\cdot\boldsymbol{X}(t) \tag{8.6}$$

with $\gamma = \gamma_A$, $\boldsymbol{v} = \boldsymbol{v}_A$ and $\gamma = \gamma_B$, $\boldsymbol{v} = \boldsymbol{v}_B$, respectively. The performance ratio for $A$ relative to $B$ is denoted $P_r(A, B)$ in the following two exercises.

EXERCISE 8.3. Show that

$$P_r(A,B) = \frac{(\gamma_A - \gamma_B)r_f + (\boldsymbol{v}_A - \boldsymbol{v}_B)\cdot(\mu - \frac{1}{2}Q(\boldsymbol{v}_A + \boldsymbol{v}_B))}{\sqrt{(\boldsymbol{v}_A - \boldsymbol{v}_B)\cdot Q(\boldsymbol{v}_A - \boldsymbol{v}_B)}}. \tag{8.7}$$

The portfolio that has maximal drift, and the portfolios that have maximal drift for given volatilities have a part of the money in the Markowitz portfolio and have the remainder in cash. The stock weights for these portfolios are given by (6.26) and (6.34).

EXERCISE 8.4. Let $A$ be the portfolio that has the weight $a$ in the Markowitz portfolio and has $1 - a$ in cash, and let the corresponding weights for $B$ be $b$ and $1 - b$, respectively. Show that

$$P_r(A, B) = \text{sign}(a - b)\frac{\omega - \frac{a+b}{2}}{\omega}\sigma_{max}. \quad (8.8)$$

Let $\sigma_A$ and $\sigma_B$ denote the volatilities of the portfolios in the above exercise. Then $\sigma_A = a\sigma_{max}/\omega$ and $\sigma_B = b\sigma_{max}/\omega$. Identity (8.8) can therefore also be written

$$P_r(A, B) = \text{sign}(\sigma_A - \sigma_B)(\sigma_{max} - \frac{\sigma_A + \sigma_B}{2}), \quad (8.9)$$

provided that $\omega > 0$.

It follows that portfolio $A$ performs better than $B$ if and only if $|a - \omega| < |b - \omega|$, or equivalently if $|\sigma_A - \sigma_{max}| < |\sigma_B - \sigma_{max}|$.

We shall now consider the case when the asset $B$ is the interest rate, $\nu_B = r_f$ and $\sigma_B = 0$, and write $\nu$ and $\sigma$ for the drift and volatility for the asset $A$. In this case we have

$$P_r = \frac{\nu - r_f}{\sigma}.$$

This ratio can thus be interpreted as the risk-adjusted expected growth over the interest rate.

A similar measure is the *Sharpe ratio,*

$$Sh = \frac{\mu - r_f}{\sigma}, \quad (8.10)$$

where $\mu$ is the expected *return*. Here we ought to use the return rate and not the continuous rate. However, the difference is negligible since

$$(e^{r_f \partial t} - 1)/\partial t - r_f \approx r_f^2 \partial t/2.$$

The Sharpe ratio is a blunt tool to judge the performance of an asset.

EXERCISE 8.5. Show that $Sh = P_r + \frac{\sigma}{2}$.

We saw in Exercise 4.15 that a high expected return can lead to a certain ruin. In the same way a high Sharpe ratio can give the same result.

Let $s$ be a given number. Consider all assets that have the Sharpe ratio $s$ — that is, for which $\mu$ and $\sigma$ satisfies $\mu = r_f + s\sigma$. It follows from Exercise 8.5 that $P_r = s - \frac{\sigma}{2}$. The probability that the asset has performed better than the interest rate at $t$ thus equals

$$\Phi((s - \frac{\sigma}{2})\sqrt{t}).$$

Regardless, the value $s$ of the Sharpe ratio this probability can be arbitrarily small if $\sigma$ is sufficiently large.

However, the difference between the Sharpe ratio and the performance ratio is not important if $\sigma$ is small compared to $Sh$.

In Exercise 8.5 we have made the assumption that the expected return is estimated by the arithmetic mean value

$$\hat{\mu} = (r_1 + \cdots + r_n)/(n\partial t),$$

where $r_1, ...., r_n$ are the returns during the different periods. If we instead replace this with the actual return,

$$r = \left(\Pi_{i=1}^{n}(1 + r_i)\right)^{1/(n\partial t)} - 1 = e^{\hat{\nu}} - 1,$$

then

$$Sh = P_r + \frac{e^{\nu} - 1 - \nu}{\sigma} \approx P_r + \frac{\nu^2}{2\sigma},$$

where the approximation is good when $\nu$ is small.

***Example 8.2.*** Data from a certain hedge fund based on 46 monthly returns gave the Sharpe ratio 1.66, the average annual return 0.1085, and the volatility 0.0446. Therefore

$$\hat{P}_r = 1.66 - \frac{0.1085 - \ln 1.1085}{0.0446} = 1.54.$$

We thus have

$$\hat{T}_{0.05} = (1.645/1.54)^2 = 1.14.$$

However, this estimate is uncertain. A 50% confidence interval for $P_r$ is

$$\hat{P}_r \pm 0.67/\sqrt{T} = (1.20, 1.88)$$

since $T = 46/12$. The time $T_{0.05}$ can therefore equally well lie outside the interval $(0.8, 1.9)$ as inside.

EXERCISE 8.6. Assume that the interest rate equals 4%. Calculate the performance ratios relative to the interest rate, and calculate the Sharpe ratios for AFGX and the portfolios A, B, and C in Section 7.3.2 using data from Table 7.9.

EXERCISE 8.7. Assume that the CAPM identity (7.15) holds. Show that the Sharpe ratios satisfy $Sh_i = \rho_i Sh_M$. Here $Sh_M$ is the Sharpe ratio of the market, and $\rho_i$ is the correlation coefficient between asset $i$ and the market.

## 8.2 RISK MEASURES

### 8.2.1 Value at Risk

The volatility is a somewhat vague risk measure: High volatility corresponds to high risk, and conversely. Value at risk is a more precise measure.

Let $R_P$ denote the return of a portfolio during a period of length $\partial t$. *Value at risk* is a number VaR such that for a given number $0 < \epsilon < 1$, the probability that $R_P < -\text{VaR}$ equals $\epsilon$. The probability for a relative loss that is greater than VaR thus equals $\epsilon$.

If we assume that the return of the portfolio is normally distributed with expectation $\mu \partial t$ and standard deviation $\sigma_P \sqrt{\partial t}$, then VaR is given by

$$\text{VaR} = z_\epsilon \sigma_P \sqrt{\partial t} - \mu \partial t.$$

With data from Tables 4.1 and 4.2 the quotient

$$\mu \partial t / (z_\epsilon \sigma_P \sqrt{\partial t})$$

varies between $3\%$ and $5\%$ when $\partial t$ is one day and $\epsilon = 5\%$. Therefore

$$\text{VaR} \approx z_\epsilon \sigma_P \sqrt{\partial t}. \tag{8.11}$$

Here follows the observed values of

$$\text{VaR}/(\sigma_P \sqrt{\partial t})$$

for *Five Stocks* during Period 1– 4. The period length is $\partial t = 1/250$ and $\epsilon = 5\%$.

| AFGX | AZN | LME | HM | SDIA | SKA |
|---|---|---|---|---|---|
| 1.6 | 1.5 | 1.5 | 1.3 | 1.3 | 1.4 |

To be compared to the quantile $z_{0.05} = 1.6$. One thus seems to be on the safe side if one uses this.

With this quantile and $\partial t = 1/250$, we have

$$VaR \approx \sigma_P/10 = 10\sigma_P\%. \tag{8.12}$$

The value of the portfolio will thus fall more than $10\sigma_P$ percent about one day out of 20 — that is, about once a month.

EXERCISE 8.8. Calculate the daily $5\%$ VaR for AFGX, and for the Portfolios A, B, and C in Section 7.3.2 using data from Table 7.9.

The actual relative frequencies became $2.8\%$, $3.6\%$, and $3.4\%$, respectively, for the portfolios A, B, and C. The normal quantile, $1.6$, thus gave values that was too pessimistic. If one replaces this with 1.4, then $100\text{VaR} \approx 9\sigma_P$, and the actual relative frequencies becomes 4.4%, 4.9%, and 4.5%, respectively, which is closer to 5%.

Portfolio (6.1) thus has the volatility (6.4). Consider more generally a portfolio that also consists of derivatives;

$$V_j(t) = a_j f_j(t, S_j(t)).$$

For a long position in a stock we thus have $f(t, s) = s$. In the case of long positions in a call or a put, $f(t, S_0)$ is given by (5.7) and (5.11), but with $T$ replaced by $T - t$. Formula (2.6) gives the function for a long position in a forward: $f(t, s) = s - e^{rt} S_0$.

The return of the portfolio thus satisfies

$$R_P(t, t + \partial t) = \sum_{j=1}^{m} a_j (f_j(t + \partial t, S_j(t + \partial t)) - f_j(t, S_j(t)))/P(t).$$

Neglecting terms of order $\partial t$, we therefore have

$$R_P(t, t + \partial t) \approx \sum_{j=1}^{m} w_j(t) R_j(t, t + \partial t).$$

Here $R_j(t, t + \partial t)$ is as in (6.3); also,

$$w_j(t) = a_j \Delta_j(t, S_j(t)) S_j(t)/P(t),$$

where $\Delta_j(t, S_j(t))$ is the delta of the derivative:

$$\Delta_j(t, s) = \frac{\partial f_j(t, s)}{\partial s}.$$

Thus this portfolio has the volatility

$$\sigma_P = \sqrt{\boldsymbol{w}(t) \cdot Q \boldsymbol{w}(t)}. \tag{8.13}$$

EXERCISE 8.9. Show that (8.13) for the three portfolios (2.9), (2.10), and (2.11) at time $t = 0$ becomes

$$\frac{\Phi(d_+(K))}{1 + p}\sigma, \quad \Big(\Phi(d_+(K)) - \Phi(d_+(H))\Big)\sigma, \quad \frac{1 - \Phi(d_+(H))}{1 - c}\sigma. \tag{8.14}$$

Here $\sigma$ is the volatility of the underlying asset.

Note that even the portfolio (2.11) has lower risk than the asset despite the fact that the asset holding is multiplied by a factor $1/(1 - c) > 1$.

***Example 8.3.*** Consider the same situation as in the Example 5.4. In this case the volatilities (8.14) are $0.63\sigma$, $0.26\sigma$, and $0.63\sigma$, respectively. The trading strategy (2.10) thus reduces the VaR in a very effective way.

**Summary 8.4.** *The probability for a relative loss during a period of length $\partial t$ that is greater than VaR approximately equals $\epsilon$. Here VaR is given by (8.11). In the particular case when $\epsilon = 5\%$ and when $\partial t =$ one day, VaR is given by (8.12).*

### 8.2.2 Downside Risk

High volatility is not necessarily the same as high risk. Assume that we have an asset that is very volatile, but that we know that the growth, $G(0, \partial t)$, always will exceed the growth of the risk-free rate, $r_f \partial t$. Such an asset can hardly be considered as being risky even if it is very uncertain with how much it will exceed the rate.

We shall define the *downside risk*, $D$, as

$$D = \sqrt{E[\min(0, G(0, \partial t) - r_f \partial t)^2]/\partial t}. \tag{8.15}$$

Here we have divided with $\partial t$ in order to give the downside risk on a yearly basis.

We thus have

$$G(0, \partial t) = \nu \partial t + \sigma \sqrt{\partial t} Y,$$

where $Y$ has expectation 0 and variance 1. (See Section 4.1.2.) Let $\delta = P_r \sqrt{\partial t}$, where $P_r = (\nu - r_f)/\sigma$ is the performance ratio relative to the interest rate. Then

$$G(0, \partial t) - r_f \partial t = \sigma \sqrt{\partial t} (\delta + Y)$$

and hence

$$D = \sigma \sqrt{F(\delta)}, \text{ where } F(\delta) = E[\min(0, \delta + Y)^2].$$

EXERCISE 8.10. Show that $F(\delta)$ decreases to 0 as $\delta$ increases to $\infty$, provided that $EY^2 < \infty$.

The downside risk will thus be low relative to the volatility when $P_r$ is large. Observe that if $P_r \neq 0$, the downside risk depends on the period length, $\partial t$. If the performance ratio is positive, then the downside risk is a decreasing function of the period length.

EXERCISE 8.11. Show that if the distribution of $Y$ is symmetrical around 0 (i.e., $-Y$ has the same distribution as $Y$), then $F(0) = 1/2$.

It is thus desirable that $D < \sigma/\sqrt{2} \approx 0.7\sigma$.

Large performance ratio is not the only way get low downside risk relative to the volatility, but this can be small even if $\delta = 0$ in the case where the distribution of $Y$ is skew. The distribution in the next exercise is chosen only to illustrate this fact. It is not intended to give a good description of the real world.

EXERCISE 8.12. Let $Y = X - 1$, where $X$ is exponentially distributed with expectation 1. Show that $F(0) = 1 - 2/e$.

In this case we thus have

$$D = \sigma \sqrt{1 - 2/e} \approx 0.5\sigma,$$

provided that $\delta = 0$.

We have seen that if $\partial t = 1/12$, a month, then the growths often are approximately normally distributed.

EXERCISE 8.13. Show that if $Y$ is standardized normally distributed, then $F(\delta) = \Psi(\delta)$, where

$$\Psi(\delta) = (1+\delta^2)\Phi(-\delta) - \delta\phi(\delta).$$

The downside risk is thus a known function of the volatility and $P_r$ in the case where the growth is normally distributed,

$$D = \sigma\sqrt{\Psi(P_r\sqrt{\partial t})}, \tag{8.16}$$

and therefore does not give any further information.

It is also seen here that the variance of the estimate of $D$ in the normally distributed case is of the order $\partial t/T$ since this holds for both $\hat{\sigma}$ and $\hat{\delta} = \hat{P}_r\sqrt{\partial t}$. See Exercise 4.5b Exercise 8.2. The estimate of $D$ thus ought to be reasonably stable.

It is customary to base the definition of the downside risk on the returns instead of the growths:

$$Dr = \sqrt{E[\min(0, R(0,\partial t) - r_f\partial t)^2]/\partial t}.$$

The mathematics becomes more tiresome, but a calculation shows that if we neglect terms of order less than $\partial t$, then we get

$$R(0,\partial t) - r_f\partial t = \sigma\sqrt{\partial t}(\delta_r + Y + U).$$

Here

$$U = \frac{\sigma\sqrt{\partial t}}{2}(Y^2-1), \quad \delta_r = Sh\sqrt{\partial t}, \quad \text{where } Sh = \frac{\nu + \frac{\sigma^2}{2} - r_f}{\sigma}$$

is the Sharpe ratio. The random variable $U$ has the expected value 0 and a standard deviation that is proportional to $\sigma\sqrt{\partial t}$. (In the case when $Y$ is normally distributed the standard deviation equals $\sigma\sqrt{\partial t/2}$.) This variable is thus negligible compared to $Y$.

Therefore

$$Dr \approx \sigma\sqrt{F(\delta_r)}.$$

***Example 8.5.*** Here is an example of a fund that has a downside risk that cannot be explained by formula (8.16). The key figures are based on 61 monthly returns.

| Sh | $\sigma$ | Dr | $\sigma\sqrt{\Psi(\delta_r)}$ | $Dr/(\sigma\sqrt{\Psi(\delta_r)})$ |
|---|---|---|---|---|
| 1.19 | 0.164 | 0.050 | 0.087 | 0.58 |

In Fig. 8.1 a normal plot for the monthly returns is shown. There are essentially two reasons for the deviation from the normal distribution: the two months when the

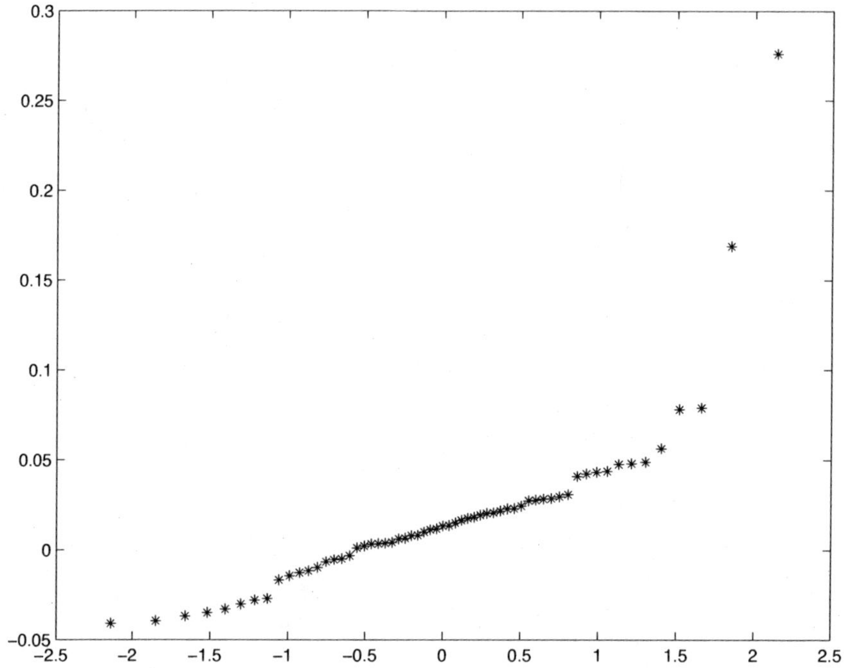


**Figure 8.1.** Normal plot for the monthly returns of the fund in Example 5.

returns became 16.88% and 27.59%, respectively. If we take away these the downside risk increases marginally to 5.09% from 5.00%. The reason for this increase is that the number of observations decreases. One the other hand, the decrease in volatility is essential, 43%, from 16.41% to 9.42%.

**Summary 8.6.** *The downside risk depends on the period length. In the case where the growth is normally distributed, it is a known function, (8.16), of the volatility and the performance ratio relative to the interest rate, and therefore it does not give any further information in this case.*

## 8.3 RISK ADJUSTMENT

Here we shall here decrease the risk in a portfolio by stepwise making small adjustments of the portfolio weights. The starting point can be an existing portfolio. Another possibility is that we want to construct a portfolio with minimal risk, but find that the portfolio that has minimal volatility or downside risk has a very large weight in a single asset. It is true that this portfolio has minimal risk measured by the history, but we will also hedge against events that the history does not catch — for example, that a company goes into liquidation. To hedge against such events, it is perhaps better to have equal weights for the different assets.

One possibility is that one starts with equal weights and then stepwise decreases the historical risk by making small adjustments of the weights. Each adjustment is in the direction that gives the largest decrease of risk. The process stops before a single asset attains a too large weight.

The volatility of the portfolio, $\sigma(\boldsymbol{v})$, is thus estimated by

$$\hat{\sigma}(\boldsymbol{v}) = \sqrt{\boldsymbol{v} \cdot \hat{Q}\boldsymbol{v}},$$

and here we shall use the definition of the downside risk, $D(\boldsymbol{v})$, that is based on the returns:

$$\hat{D}(\boldsymbol{v})^2 = \frac{1}{n\partial t}\sum_{j=1}^{n} \min(0, \boldsymbol{v} \cdot \boldsymbol{R}_j - r_f \partial t)^2.$$

Here $\boldsymbol{R}_j = (R_{1\,j}, ..., R_{m\,j})$ are the returns of the assets of the portfolio during period $j$, $j = 1, ..., n$.

We shall make a small adjustment $\boldsymbol{v} \to \boldsymbol{v} + \partial\boldsymbol{v}$ of the weights. Even $\boldsymbol{v} + \partial\boldsymbol{v}$ shall be weights and hence we must have

$$\partial\boldsymbol{v} \in L = \{\boldsymbol{x} \in \mathbb{R}^m; \boldsymbol{x} \cdot \mathbf{1} = 0\}.$$

$L$ is thus a linear subspace of $\mathbb{R}^m$ of dimension $m - 1$.

EXERCISE 8.14. Show that

$$L = (I - \frac{1}{m}J)\mathbb{R}^m,$$

that is, that $\boldsymbol{x} \in L$ if and only if

$$\boldsymbol{x} = (I - \frac{1}{m}J)\boldsymbol{y}$$

for some $\boldsymbol{y} \in \mathbb{R}^m$. Here $J$ is the $m \times m$ matrix having all elements equal to 1.

Let $F$ be a regular function of the weights. We have

$$F(\boldsymbol{v} + \partial\boldsymbol{v}) = F(\boldsymbol{v}) + \partial\boldsymbol{v} \cdot \nabla F(\boldsymbol{v}) + O(|\partial\boldsymbol{v}|^2) =$$

$$F(\boldsymbol{v}) + \partial\boldsymbol{v} \cdot (I - \frac{1}{m}J)\nabla F(\boldsymbol{v}) + O(|\partial\boldsymbol{v}|^2)$$

for $\partial\boldsymbol{v} \in L$.

We shall minimize this expression when $\partial\boldsymbol{v}$ is small. More precisely over all, $\partial\boldsymbol{v} \in L$ for which $|\partial\boldsymbol{v}| = \epsilon$, where $\epsilon$ is a small number. If we neglect terms of the order $\epsilon^2$, this is the same as to minimize

$$\partial\boldsymbol{v} \cdot (I - \frac{1}{m}J)\nabla F(\boldsymbol{v}).$$

This minimum is attained for

$$\partial \boldsymbol{v} = -h(I - \frac{1}{m}J)\nabla F(\boldsymbol{v}),$$

where $h > 0$ is chosen such that $|\partial \boldsymbol{v}| = \epsilon$.

EXERCISE 8.15. Show that $(I - \frac{1}{m}J)\nabla F(\boldsymbol{v})$

$$= \hat{Q}\boldsymbol{v} - \frac{\mathbf{1} \cdot \hat{Q}\boldsymbol{v}}{m}\mathbf{1}$$

when $F(\boldsymbol{v}) = \frac{1}{2}\hat{\sigma}(\boldsymbol{v})^2$ and

$$= \frac{1}{n\partial t}\sum_{j=1}^{n} \min(0, \boldsymbol{v} \cdot \boldsymbol{R}_j - r_f \partial t)\Big(\boldsymbol{R}_j - \frac{\mathbf{1} \cdot \boldsymbol{R}_j}{m}\mathbf{1}\Big)$$

when $F(\boldsymbol{v}) = \frac{1}{2}\hat{D}(\boldsymbol{v})^2$.

This method can also be used to calculate the portfolios that have minimal volatility or downside risk.

Table 8.1 is based on monthly data during 47 months. The portfolios consists of AFGX and two hedge funds, E and F. The three first figures in each row are the weights, and the last is the portfolio volatility these weights give. The unit is %. The last row gives the weights and volatility of the Minimum variance portfolio. We start at row 5 with equal weights and go in the direction of the gradient towards the Minimum variance portfolio. A few examples of portfolios we will pass are given in the rows 6, 7 and 8.

**Table 8.1.** Decrease of the volatility along the gradient.

| AFGX | E | F | $\sigma$ |
|---|---|---|---|
| 100 | 0 | 0 | 25.7 |
| 0 | 100 | 0 | 7.0 |
| 0 | 0 | 100 | 14.7 |
| 33 | 33 | 33 | 7.8 |
| 25 | 40 | 36 | 6.7 |
| 19 | 50 | 31 | 6.1 |
| 17 | 61 | 23 | 5.8 |
| 15 | 72 | 13 | 5.6 |

Table 8.2 is as Table 8.1 but with the difference that the volatility is replaced by the downside risk. When calculating this, we have neglected that the interest rate has varied during the period, and we used the constant monthly rate $r_f = 0.03/12$.

The two methods thus give similar results despite the fact that AFGX during this period has a downside risk that is very high relative to the volatility.

**Table 8.2.** Decrease of the downside risk along the gradient.

| AFGX | E | F | $D$ |
|---|---|---|---|
| 100 | 0 | 0 | 20.2 |
| 0 | 100 | 0 | 3.6 |
| 0 | 0 | 100 | 8.1 |
| 33 | 33 | 33 | 4.7 |
| 23 | 40 | 37 | 3.6 |
| 16 | 50 | 34 | 3.1 |
| 13 | 60 | 27 | 2.8 |
| 12 | 70 | 18 | 2.7 |

# CHAPTER 9

# SIMPLE COVARIATION

In this chapter we shall consider two models with simplified covariation structure: equal correlations and multiplicative correlations. There are three reasons for this.

One reason is that we will get simple and lucid explicit expressions for portfolio weights and other quantities. This will give a better understanding of what is important.

Another reason is that the number of parameters in the covariance matrix to be estimated is drastically reduced, and this seems to make the estimates more stable as the random error decreases. It is true that the systematical error increases, but we shall see that we cannot tell whether these methods or the one with a general covariance matrix will give the best result for *Five Stocks*. Therefore the simple models are probably to prefer when plenty of data are not available. This is the case if there are stocks with a short history in the portfolio.

A third reason is that these models will give a simple solution to the problem to find optimal portfolio weights under the constraint that these must not be negative.

*Mathematical Asset Management.* By Thomas Höglund

We shall, at the cost of some repetition, treat the two models separately.

## 9.1 EQUAL CORRELATIONS

In this case

$$\sigma_{i,i} = \sigma_i^2 \text{ and } \sigma_{i,j} = \sigma_i \sigma_j \rho \quad \text{for } i \neq j. \tag{9.1}$$

The number of parameters is thus $m+1$, to be compared to $m(m+1)/2$ for a general covariance matrix.

The correlations for *Five Stocks* are shown in Table 4.9. These numbers are of similar size. They have the mean value $0.28$ and the standard deviation $0.06$. Therefore the model (9.1) ought to agree fairly well with the reality in this case. On the other hand, this model is not realistic if the portfolio contains an asset that is negatively correlated with the others.

### 9.1.1 Matrix Calculations

We shall start with the case $\sigma_i = 1$ for $i = 1, ..., m$, and write $Q_0$ for the covariance matrix in this case.

Let $I$ stand for the identity matrix, and let $J$ stand for the matrix having all elements equal to 1. Then

$$Q_0 = (1-\rho)I + \rho J \text{ and hence } Q_0^{-1} = (1-\rho)^{-1}(I + \frac{\rho}{1-\rho}J)^{-1}.$$

EXERCISE 9.1.

(a) Show that

$$\boldsymbol{x} \cdot Q_0 \boldsymbol{x} = m(1-\rho)v(\boldsymbol{x}) + m(1+(m-1)\rho)\bar{x}^2,$$

where $v(\boldsymbol{x}) = \frac{1}{m}\sum_{i=1}^m (x_i - \bar{x})^2$.

(b) Show that $Q_0$ (and hence also $Q$) is strictly positive definite if and only if

$$-\frac{1}{m-1} < \rho < 1. \tag{9.2}$$

We shall assume that condition (9.2) is satisfied.

EXERCISE 9.2.

(a) Show that $J^2 = mJ$.

(b) Verify that

$$Q_0^{-1} = (1-\rho)^{-1}(I - \kappa J)$$

where

$$\kappa = \frac{\rho}{1+(m-1)\rho}. \tag{9.3}$$

We now release the restriction $\sigma_i = 1$ and consider general volatilities. Let $S$ denote the diagonal matrix with the elements $\sigma_1, ..., \sigma_m$. Then

$$Q = SQ_0S \text{ and hence } Q^{-1} = S^{-1}Q_0^{-1}S^{-1}.$$

EXERCISE 9.3. Show that

$$(Q^{-1}\boldsymbol{x})_i = \frac{1}{(1-\rho)\sigma_i}\Big(\frac{x_i}{\sigma_i} - \kappa\sum_{j=1}^{m}\frac{x_j}{\sigma_j}\Big). \tag{9.4}$$

### 9.1.2 Optimal Portfolios

***The Minimum Variance Portfolio*** It follows from (6.9) and (9.4) that the Minimum variance portfolio has the weights

$$v_i = \frac{\sigma_{min}^2}{(1-\rho)\sigma_i}\Big(\frac{1}{\sigma_i} - \kappa\sum_{j=1}^{m}\frac{1}{\sigma_j}\Big). \tag{9.5}$$

A way to calculate these weights and the volatility is given in the next exercise.

EXERCISE 9.4 Put

$$w_i = \frac{1}{\sigma_i}\Big(\frac{1}{\sigma_i} - \kappa\sum_{j=1}^{m}\frac{1}{\sigma_j}\Big).$$

Show that $v_i = w_i/\sum_{j=1}^m w_j$ and $\sigma_{min}^2 = (1-\rho)/\sum_{j=1}^m w_j$.

***The Maximal Drift Portfolio*** The weights (6.26) and the variance (6.27) take the form

$$v_{max}(i) = \frac{1}{(1-\rho)\sigma_i}\Big(\frac{\mu_i - r_f}{\sigma_i} - \kappa\sum_{j=1}^{m}\frac{\mu_j - r_f}{\sigma_i}\Big) \tag{9.6}$$

and

$$\sigma_{max}^2 = \frac{1}{1-\rho}\Big(\sum_{i=1}^{m}\Big(\frac{\mu_i - r_f}{\sigma_i}\Big)^2 - \kappa\Big(\sum_{i=1}^{m}\frac{\mu_i - r_f}{\sigma_i}\Big)^2\Big). \tag{9.7}$$

The weights of the Markowitz portfolio are calculated from (9.6);

$\boldsymbol{v}_{Mz} = \boldsymbol{v}_{max}/\boldsymbol{1}\cdot\boldsymbol{v}_{max}$. These are thus simple expressions of the Sharpe ratios (8.10).

### *The Beta Portfolio*

EXERCISE 9.5. Show that

$$\sigma_M^2 = \rho\bar{\sigma}^2 + (1-\rho)\sum_{j=1}^{N}(p_j\sigma_j)^2$$

and

$$\sigma_M^2\beta_i = \sigma_i(\rho\bar{\sigma} + (1-\rho)p_i\sigma_i), \tag{9.8}$$

where

$$\bar{\sigma} = \sum_{j=1}^{N} p_j\sigma_j.$$

EXERCISE 9.6. Show that the weights of the Beta portfolio are proportional to

$$w_i = p_i + \frac{\bar{\sigma}'_m\kappa}{\sigma_i}, \tag{9.9}$$

where $\bar{\sigma}'_m = \sum_{i>m} p_i\sigma_i$.

The Beta portfolio thus has positive weights in this case.

When $\rho = 0$ the weights are proportional to the index weights, $v_\beta(i) = p_i/s_m$, where $s_m = p_1 + \cdots + p_m$.

EXERCISE 9.7. Assume that $\rho > 0$ and $m < N$. Show that $v_\beta(i) < p_i/s_m$ if and only if

$$\frac{1/\sigma_i}{1/\sigma_1 + \cdots + 1/\sigma_m} < \frac{p_i}{p_1 + \cdots + p_m}.$$

Thus, when $\rho > 0$ stocks with high volatility and heavy weight in the index will be weighted down relative to the index.

It is not necessary to know the volatilities for all stocks in the index to calculate $\bar{\sigma}$, but it suffices to use the ones with most heavy weights. The first identity in Exercise 9.5 can also be written

$$\bar{\sigma}^2 = \frac{\sigma_M^2}{\rho}(1-S), \quad \text{where } S = \sum_{j=1}^{N}(1-\rho)p_j^2\sigma_j^2/\sigma_M^2. \tag{9.10}$$

EXERCISE 9.8. Show that the contribution to the sum $S$ from assets that satisfies $(1-\rho)p_j\sigma_j^2/\sigma_M^2 < \epsilon$ is $< \epsilon$.

We thus have the inequality

$$\bar{\sigma} < \frac{\sigma_M}{\sqrt{\rho}}, \tag{9.11}$$

and this upper bound is a good approximation in the cases where no asset has heavy weight in the index.

### 9.1.3 Comparison with the General Model

Here we shall use *Five Stocks* and compare the weights of the Minimum variance and the Beta portfolio calculated with a general covariance matrix, with the weights we obtain assuming that the correlations are equal.

***The Minimum Variance Portfolio*** We shall use data from Period 1– 4. As an estimate of $\rho$, we shall use the mean value of the correlations in Table 4.9: $\hat{\rho} = 0.284$.

EXERCISE 9.9. Calculate the volatility and the weights for the Minimum variance portfolio using this $\rho$ and the volatilities in Table 4.1.

The differences between these weights and the weights in the last row in Table 6.1 are $-0.04$, $-0.01$, $-0.02$, $0.05$, $0.03$. The latter portfolio has the volatility $0.223$. The mean deviation between the two estimates of the weights equals

$$d' = \sqrt{\frac{(-0.04)^2 + \cdots + 0.03^2}{5}} = 0.033.$$

Table 6.4 indicates that one can expect that the estimation error approximately equals $\sqrt{2} \times 0.02 \times 1.8 \approx 0.05$. (The factor 1.8 is used since the observed values are about 1.8 times the theoretical values.) It follows that we cannot decide which of the two estimates that is closest to the Minimum variance portfolio. The model with equal correlations thus give a very good result in this case.

If we simplify the model further and assume that $\rho = 0$, we get the weights $(0.27, 0.11, 0.15, 0.12, 0.36)$ and $d' = 0.087$. This estimate can therefore possibly be distinguished from the other two. The three portfolios have however a similar character which is seen if one ranks the portfolio weights.

***The Beta Portfolio*** The parameters were estimated with data from Period 1– 3. The index weights were taken from the end of Period 3: October 29, 1999. These are in the first row in Table 9.1. The correlations were estimated by the mean value of the correlations, $\hat{\rho} = 0.37$, and $\bar{\sigma}$ were estimated by $0.32$. See below. The weights for the Beta portfolio with equal correlations is in the fourth row in Table 9.1. For a comparison, also the weights of the portfolio that has the weights proportional to the index weights, $\boldsymbol{v}_{\beta}^{prop}$, have been calculated. The distance between these portfolio weights and $\boldsymbol{v}_{\beta}$, the weights of the Beta portfolio with a general covariance matrix, are given in the rightmost column. Also the weighted mean value $\bar{\sigma}_m = \sum_{j=1}^{m} p_j \sigma_j / s_m$ is given. The unit is %.

**Table 9.1.** The Beta portfolios

| | AZN | LME | HM | SDIA | SKA | Dist. |
|---|---|---|---|---|---|---|
| Index weights | 5.90 | 16.15 | 6.16 | 3.17 | 1.16 | – |
| $v_\beta^{prop}$ | 18 | 50 | 19 | 10 | 4 | 7.7 |
| $v_\beta$ | 20 | 40 | 11 | 16 | 13 | 0 |
| $v_\beta^{equal}$ | 27 | 31 | 16 | 15 | 11 | 4.8 |
| Volatilities | 31 | 45 | 36 | 44 | 28 | $\bar{\sigma}_m = 40$ |

The upper bound (9.11) equals $0.3482$, but this will probably give a value for $\bar{\sigma}$ that is too large since there are stocks with heavy weights in this case.

The contribution from *Five Stocks* to the sum $S$ in (9.10) is given in Table 9.2.

**Table 9.2.** The contribution from *Five Stocks*

| AZN | LME | HM | SDIA | SKA | Sum |
|---|---|---|---|---|---|
| 0.0047 | 0.0764 | 0.0070 | 0.0027 | 0.0001 | 0.0910 |

The total weight of these companies in index equals $0.33$. If one replaces $S$ with only the contribution from LME, the upper bound will sink to $0.3346$. If one uses all five stocks, one gets $0.3320$. Except for HM and AZN, there are a few other companies with heavy weights in the index, but that are not in our portfolio. Therefore $\bar{\sigma} = 0.32$.

We shall now argue as for the Minimum variance portfolio. Table 7.5 indicates that the distance ought to be approximately $\sqrt{2} \times 0.03 \times 1$ if the period length had been 1024 days, but it is $768 = \frac{3}{4} \times 1024$ days. Therefore the random error ought to be approximately $\sqrt{4/3} \times \sqrt{2} \times 0.03 \approx 5\%$. Therefore we cannot either in this case decide which of the estimates that is closest to the truth (i.e., the Beta portfolio).

In Fig. 9.1, correlations are shown together with AFGX and the Beta portfolio with a general covariance matrix. The first is dashed and lies between the other two that are unbroken.

### 9.1.4 Positive Weights

Here we shall give an algorithm that will solve the problem to determine the weights of the portfolios that have minimal variance or maximal drift under the constraint that the weights are positive. More precisely, we shall give an algorithm that solves (6.55). The optimal set that solves (6.55), and for which all weights are strictly positive, will be denoted $I_*$. It will follow from the construction below that this set is unique.

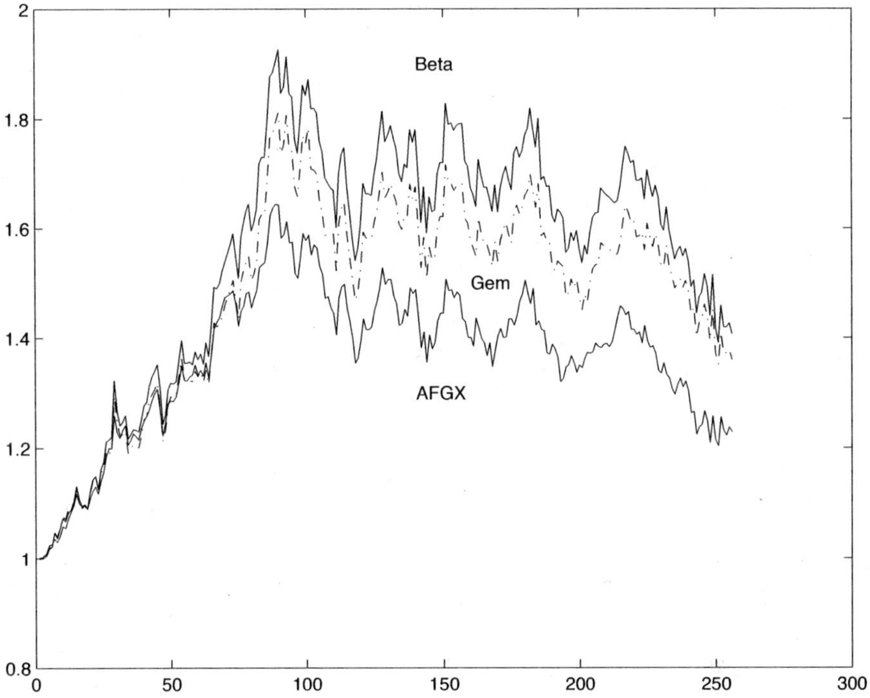


**Figure 9.1.** The development of the Beta portfolio with equal correlations during Period 4.

Write $[1:m] = \{1, 2, ..., m\}$. We have for $I \subset [1:m]$

$$(Q_I^{-1}\boldsymbol{u}_I)_i = \frac{1}{\sigma_i(1-\rho)}\big(x_i - \kappa(I)\sum_{j\in I} x_j\big) \quad \text{for } i \in I \tag{9.12}$$

and

$$\boldsymbol{u}_I \cdot Q_I^{-1}\boldsymbol{u}_I = \frac{1}{1-\rho}\Big(\sum_{j\in I} x_j^2 - \kappa(I)\big(\sum_{j\in I} x_j\big)^2\Big).$$

Here

$$x_i = \frac{u_i}{\sigma_i}, \quad \kappa(I) = \frac{\rho}{1-\rho+|I|\rho},$$

and $|I|$ stands for the number of elements in $I$.

The weights for the Maximal drift portfolio are then given by (9.12) with $x_i = (\mu_i - r_f)/\sigma_i$, whereas the weights for the Minimum variance portfolio are proportional to (9.12) with $x_i = 1/\sigma_i$.

***Example 9.1.*** Uncorrelated Returns.

In this case we shall maximize
$\sum_{i \in I} x_i$ subject to $x_i > 0$ for $i \in I$. It therefore follows that $I_* = \{i; \mu_i > r_f\}$ for the maximal drift portfolio and that $I_* = \{1, 2, ..., m\}$ for the minimum variance portfolio. □

Here we shall describe how to construct the weights. The proofs will be given in the next section on multiplicative correlations. These proofs will formally cover equal correlations only when $\rho \geq 0$, but they can easily be modified to cover even the case $-\frac{1}{m-1} < \rho < 0$.

It turns out that the obvious algorithm gives the right answer for equal correlations.

**Algorithm 9.2.**

*1. Start with* $I = \{1, ..., m\}$.

*2. Calculate (9.12), and let* $J$ *be the set of* $i$ *for which these are strictly positive.*

*3. Stop if* $J = I$. *Otherwise let* $I = J$ *and go to 2.*

It may happen that this algorithm has to be repeated.

EXERCISE 9.10. Show this by constructing an example.

*Hint:* It suffices to consider the case when $m = 3$.

The optimal set $I_*$ is thus of the form

$$I_* = \{i \in [1 : m]; x_i > \xi_*\}, \tag{9.13}$$

where

$$\xi_* = \kappa(I_*) \sum_{j \in I_*} x_j.$$

Another possibility is to first determine $\xi_*$ and then determine $I_*$ from (9.13).

**Lemma 9.3.** *Put*

$$f(\xi) = (1 - \rho)\xi + \rho \sum_{i=1}^{m} \min(0, \xi - x_i).$$

*The function* $f$ *has a unique zero,* $\xi_*$.

Note that $f$ is a piecewise linear and continuous function that is strictly increasing since

$$f'(\xi) = 1 - \rho + \rho|\{i; x_i > \xi\}| \tag{9.14}$$

except when $\xi = x_i$.

The value $\xi_*$ can be calculated by Newtons method:

$$\xi_n = \xi_{n-1} - f(\xi_{n-1})/f'(\xi_{n-1}). \tag{9.15}$$

Here $f'(\xi)$ is defined by the expression to the right in (9.14) when $\xi = x_k$. The exact value of $\xi_*$ is reached after a finite number of steps since $f$ is piecewise linear.

We can use any initial value $\xi_0$. However, if $f(x_{min}) > 0$, then $\xi_* < x_i$ for all $i \in [1:m]$, and hence it is unnecessary to start with a value $\xi_0 < x_{min}$. We therefore also have the following algorithm.

**Algorithm 9.4.**

*1. Put* $\xi_0 = min_{i\in[1:m]} x_i$.

*2. If* $f(\xi_0) > 0$, *then* $I_* = [1:m]$. *Otherwise use the recursion (9.15) to find the first* $n \geq 0$ *for which* $f(\xi_n) = 0$.

*3. The optimal set* $I_*$ *is given by (9.13) with* $\xi_* = \xi_n$.

In Exercise 9.9 we have seen that the Minimum variance portfolio for *Five Stocks* has positive weights with data from Period 1– 4. This rarely holds for larger portfolios, however.

EXERCISE 9.11. Consider a portfolio that consists of 40 stocks. The stocks consist of 5 groups of 8 stocks each. Each stock within each group has the same volatility, and the volatilities of the 5 groups are as for the stocks in *Five Stocks* with data from Period 1– 4. The common correlation equals 0.28.

(a) Determine the weights and the volatility of the Minimum variance portfolio when negative weights are allowed.

(b) Do the same when negative weights are not allowed.

(c) What is the volatility of the portfolio that has equal weights?

If the groups have size $g$ instead of 8, it can be shown that the volatility of portfolio (a) tends to zero as $g$ tends to infinity. The stocks in the groups SKA and AZN will have positive weights and the other negative. The volatility of portfolio (b) will tend to 14.3%, and the weights will equal 0 in all groups except SKA. The volatility of portfolio (c) will tend to 21.0%.

## 9.2 MULTIPLICATIVE CORRELATIONS

In this case, $\sigma_{i,i} = \sigma_i^2$ and $\sigma_{i,j} = \sigma_i \sigma_j \rho_i \rho_j$ for $i \neq j$. The correlation coefficients, $\rho_{i,j}$, are thus assumed to have the form

$$\rho_{i,j} = \rho_i \rho_j,$$

where $\rho_1, ..., \rho_m$ are $m$ parameters. The number of parameters is thus $2m$ in this case. Equal correlations is the special case $\rho_1 = \cdots = \rho_m$, provided that $\rho \geq 0$.

### 9.2.1 Uniqueness of the Parameters

Let $\boldsymbol{\delta} = (\delta_1, ..., \delta_m)$, where $\delta_1, ..., \delta_m$ are given real numbers. We shall find all solutions $\boldsymbol{\rho} = (\rho_1, ..., \rho_m)$ of the equation

$$\rho_i \rho_j = \delta_i \delta_j, \quad 1 \leq i < j \leq m. \tag{9.16}$$

The equation obviously has the solutions $\boldsymbol{\rho} = \boldsymbol{\delta}$ and $\boldsymbol{\rho} = -\boldsymbol{\delta}$. If at most two of the $\delta_i$ are nonzero, then there are more solutions than these but otherwise not.

**Lemma 9.5.** *Assume that $m \geq 3$, and that at least 3 of the $\delta_i$ are nonzero. Equation (9.16) has exactly the two solutions $\boldsymbol{\rho} = \boldsymbol{\delta}$ and $\boldsymbol{\rho} = -\boldsymbol{\delta}$.*

*Proof.* Let $r \geq 3$ be the number of nonzero parameters. Assume that $\delta_1 \neq 0$, ..., $\delta_r \neq 0$ and that $\delta_i = 0$ for $r < i \leq m$. Then $\rho_1 \neq 0$, ..., $\rho_r \neq 0$, and $\rho_i = 0$ for $r < i \leq m$. We have $\rho_i = \delta_i \delta_1 / \rho_1$ for $i = 2, ..., r$, and hence $\rho_2 \rho_3 = \delta_2 \delta_3 (\delta_1 / \rho_1)^2$. Therefore $\delta_1 / \rho_1 = 1$ or $\delta_1 / \rho_1 = -1$. In the first case $\boldsymbol{\rho} = \boldsymbol{\delta}$, and in the second $\boldsymbol{\rho} = -\boldsymbol{\delta}$. □

### 9.2.2 Matrix Calculations

EXERCISE 9.12. Show that:

(a) If $Q$ is strictly positive definite, then $|\rho_i| < 1$ for all except possibly one $i$.

(b) If $|\rho_i| < 1$ for all $i$, then $Q$ is strictly positive definite.

The exact condition is that $Q$ is strictly positive definite if and only if $|\rho_i| < 1$ for all except possibly one $i$, and if for example $|\rho_m| \geq 1$, then

$$(\rho_m^2 - 1) \sum_{k=1}^{m-1} \frac{\rho_k^2}{1 - \rho_k^2} < 1.$$

We shall here assume that $|\rho_i| < 1$ for all $i$.

Let $\Delta$ stand for the diagonal matrix with the diagonal elements

$$\sigma_i \sqrt{1 - \rho_i^2}, \; i = 1, ..., m,$$

and let $D$ stand for the diagonal matrix with diagonal elements

$$\rho_i / \sqrt{1 - \rho_i^2}, \quad i = 1, ..., m.$$

Then

$$Q = \Delta(I + DJD)\Delta,$$

and hence

$$Q^{-1} = \Delta^{-1}\big(I + DJD\big)^{-1}\Delta^{-1}.$$

EXERCISE 9.13. Let $d_i = \rho_i^2/(1-\rho_i^2)$.

(a) Show that $JD^2J = dJ$, where $d = d_1 + \cdots + d_m$

(b) Verify that

$$(I + DJD)^{-1} = I - \frac{1}{1+d}DJD.$$

EXERCISE 9.14. Show that

$$(Q^{-1}\boldsymbol{x})_i = \frac{1}{\sigma_i(1-\rho_i^2)}\Big(\frac{x_i}{\sigma_i} - \frac{\rho_i}{1+d}\sum_{j=1}^{m}\frac{\rho_j}{1-\rho_j^2}\frac{x_j}{\sigma_j}\Big). \tag{9.17}$$

### 9.2.3 Parameter Estimation

In order to get a consistent estimate of the parameters $\rho_i$, we can minimize the quadratic form

$$\sum_{i\neq j}(\rho_i\rho_j - \hat{\rho}_{i,j})^2.$$

This gives the equations

$$\sum_{j;j\neq i}(\rho_i\rho_j - \hat{\rho}_{i,j})\rho_j = 0,$$

that is,

$$\rho_i = \frac{\sum_{j;j\neq i}\hat{\rho}_{i,j}\rho_j}{\sum_{j;j\neq i}\rho_j^2}.$$

This system can be solved by iteration: Start with an initial value $\rho_i(0)$, then calculate $\rho_i(1)$, $\rho_i(2)$,... from the formula

$$\rho_i(k) = \frac{\sum_{j;j\neq i}\hat{\rho}_{i,j}\rho_j(k-1)}{\sum_{j;j\neq i}\rho_j(k-1)^2}.$$

EXERCISE 9.15. Use the data in Table 4.9 to estimate $\rho_1$, ..., $\rho_5$ for *Five Stocks*. Then calculate the estimated correlations and compare with Table 4.9.

### 9.2.4 Optimal Portfolios

The weights and variances are obtained from (9.17). Thus the weights of the minimum variance portfolio are proportional to

$$\frac{d_i}{\tau_i}(\frac{1}{\tau_i} - \frac{1}{1+d}\sum_{j=1}^{m}\frac{d_j}{\tau_j}) \text{ if } \rho_i \neq 0, \ \frac{1}{\sigma_i^2} \text{ if } \rho_i = 0.$$

Here $\tau_i = \rho_i\sigma_i$ and $d_j/\tau_j = 0$ if $\rho_j = 0$.

EXERCISE 9.16. Use the data for *Five Stocks* from Period 1– 4 to determine the weights of the Minimum variance portfolio. (The volatilities are in Table 4.1, and the $\rho_i$ are in Exercise 9.15.)

Compare with equal correlations (Exercise 9.9) and general correlations (Table 6.1).

As another example, we shall consider the Beta portfolio and write down the formulas that correspond to those in Exercises 9.5 and 9.6.

$$\sigma_M^2 = \tau^2 + \sum_{j=1}^{N}(1-\rho_j^2)(p_j\sigma_j)^2 \tag{9.18}$$

and

$$\sigma_M^2\beta_i = \sigma_i(\rho_i\tau + (1-\rho_i^2)p_i\sigma_i),$$

where $\tau = \sum_{j=1}^{N} p_j\rho_j\sigma_j$.

The weights of the Beta portfolio are proportional to

$$p_i + \frac{\tau_m'}{\sigma_i}\frac{\rho_i}{1-\rho_i^2}\frac{1}{1+d} \quad \text{for } i = 1, ..., m,$$

where

$$\tau_m' = \sum_{j>m} p_j\tau_j.$$

The following argument is sometimes used to motivate multiplicative correlations: Define $e_i$ by

$$R_i = \alpha_i + \beta_i R_M + e_i,$$

where $\alpha_i$ is chosen so that $Ee_i = 0$. Then $e_i$ and $R_i$ are uncorrelated, and hence

$$\sigma_{ij} = \beta_i\beta_j\sigma_M^2 + \text{Cov}(e_i, e_J). \tag{9.19}$$

Assume that $e_i$ and $e_j$ are uncorrelated for $i \neq j$, then

$$\sigma_{ij} = \beta_i\beta_j\sigma_M^2. \tag{9.20}$$

That's it. Or is it?

**Lemma 9.6.** *Assume that $\sigma_M > 0$. The following three statements are equivalent.*

*(a) $e_i$ and $e_j$ are uncorrelated for $i \neq j$.*

*(b) The identity (9.20) holds for $i \neq j$.*

*(c) There are constants $a_1, ..., a_N$ and $b_1, ..., b_N$ such that*

$$R_i = a_i + b_i R_M \tag{9.21}$$

*for $i = 1, ..., N$.*

*Proof.* It follows from (9.19) that (a) and (b) are equivalent.

Assume that (b) holds. If $\sigma_i = 0$, then (9.21) holds with $b_i = 0$. Otherwise put $\rho_i = \beta_i \sigma_M / \sigma_i$. Then $\rho_i = \rho(R_i, R_M)$, and hence $|\rho_i| \leq 1$. We have

$$\sum p_i \sigma_i \rho_i = \sum p_i \beta_i \sigma_M = \sigma_M,$$

where the sum is over those $i$ for which $\sigma_i > 0$. It follows from (9.18) that

$$\sum p_i^2 \sigma_i^2 (1 - \rho_i^2) = 0,$$

and hence that $|\rho_i| = 1$ when $\sigma_i > 0$. Identity (9.21) thus holds with $b_i > 0$ if $\rho_i = 1$, and it holds with $b_i < 0$ if $\rho_i = -1$.

Assume that (c) holds. Then $\sigma_{ij} = b_i b_j \sigma_M^2$ and $\beta_i = b_i$. □

Identity (9.20) thus holds only on a truly degenerate market. This, however, does not prevent us from using multiplicative correlations. It is the interpretation of $\rho_i$ as the correlation between $R_i$ and $R_M$ that is false.

### 9.2.5 Positive Weights

Let $I \subset [1 : m]$. It follows from Exercise 9.14 that

$$(Q_I^{-1} \boldsymbol{u}_I)_i = \frac{1}{\sigma_i (1 - \rho_i^2)} \left(x_i - \frac{\rho_i}{1 + d(I)} \sum_{j \in I} \frac{\rho_j}{1 - \rho_j^2} x_j\right) \text{ for } i \in I. \tag{9.22}$$

and hence

$$\boldsymbol{u}_I \cdot Q_I^{-1} \boldsymbol{u}_I = \sum_{i \subset I} \frac{x_i}{1 - \rho_i^2} \left(x_i - \frac{\rho_i}{1 + d(I)} \sum_{j \in I} \frac{\rho_j}{1 - \rho_j^2} x_j\right) = \tag{9.23}$$

$$\sum_{i \in I} \frac{1}{1 - \rho_i^2} x_i^2 - \frac{1}{1 + d(I)} \left(\sum_{j \in I} \frac{\rho_j}{1 - \rho_j^2} x_j\right)^2.$$

Here

$$x_i = \frac{u_i}{\sigma_i}, \text{ and } d(I) = \sum_{i \in I} d_i.$$

We shall maximize (9.23) over $I$ under the constraints that (9.22) are positive.

We need a lemma to describe the optimal set.

**Lemma 9.7.** *Put*

$$f(\xi) = \xi + \sum_{i=1}^{m} \frac{\rho_i}{1-\rho_i^2} \min(0, \rho_i \xi - x_i).$$

*The function $f$ is continuous and strictly increasing and has a unique zero, $\xi_*$, satisfying*

$$sign(\xi_*) = sign( \sum_{i; x_i > 0} \frac{\rho_i}{1-\rho_i^2} x_i).$$

*Proof.* The function $f$ is strictly increasing since

$$f'(\xi) = 1 + \sum_{i; x_i > \rho_i \xi} d_i \tag{9.24}$$

except when $\xi = x_i/\rho_i$ and $\rho_i \neq 0$. The existence of the zero now follows from $f(-\infty) = -\infty$ and $f(\infty) = \infty$, whereas the statement about the sign follows from the identity

$$f(0) = - \sum_{i; x_i > 0} \frac{\rho_i}{1-\rho_i^2} x_i.$$

□

**Theorem 9.8.** *Let*

$$I_* = \{i \in [1:m]; x_i > \rho_i \xi_*\}, \tag{9.25}$$

*where $\xi_*$ is as in Lemma 9.7.*

*The maximum of the quadratic form (9.23) under the constraints that (9.22) are positive is attained for $I$ if and only if $I = I_*$.*

Note that

$$\xi_* = \frac{\sum_{i \in I_*} \frac{\rho_i}{1-\rho_i^2} x_i}{1 + d(I_*)}. \tag{9.26}$$

The value $\xi_*$ can be calculated by Newtons method:

$$\xi_n = \xi_{n-1} - f(\xi_{n-1})/f'(\xi_{n-1}). \tag{9.27}$$

Here $f'(\xi)$ is defined by the expression to the right in (9.24) when $\rho_k \xi = x_k$. The exact value of $\xi_*$ is reached after a finite number of steps since $f$ is piecewise linear.

We can choose any initial value $\xi_0$. However, if $\rho_i > 0$ for all $i$, then it is unnecessary to choose an initial value less than $\xi_0 = min_{i\in[1:m]} \frac{x_i}{\rho_i}$, but we can proceed as in Algorithm 9.4.

The following algorithm covers the case with arbitrary signs of the $\rho_i$.

**Algorithm 9.9.**

*1. Let $\xi_0$ be any real number, $\xi_0 = 0$, say.*

*2. Use the recursion (9.27) to find the first $n \geq 0$ for which $f(\xi_n) = 0$.*

*3. The optimal set $I_*$ is given by (9.25) with $\xi_* = \xi_n$.*

We have the following corollary to Theorem 9.8.

**Corollary 9.10.** *Assume that $\rho_i \geq 0$ for all $i$. Then the algorithm below yields the optimal set $I_*$.*

**Algorithm 9.11.**

*1. Start with $I = [1 : m]$.*

*2. Calculate (9.22). Let $J$ be the set of $i \in I$ for which these are strictly positive.*

*3. Stop if $J = I$. Otherwise put $I = J$, and go to 2.*

Note that if $I^0 = \{i \in [1 : m]; \rho_i = 0\}$, then $(Q_I^{-1}\boldsymbol{u}_I)_i = (Q_{I\setminus I^0}^{-1}\boldsymbol{u}_{I\setminus I^0})_i$ for $i \in I \setminus I^0$, and $(Q_I^{-1}\boldsymbol{u}_I)_i = (Q_{I^0}^{-1}\boldsymbol{u}_{I^0})_i = \frac{x_i}{\sigma_i}$ for $i \in I^0$. Therefore, the contribution from $I^0$ to the optimal set is $\{i \in I^0; x_i > 0\}$, in agreement with the theorem.

For the remaining indices we have $\rho_i \neq 0$, and we shall therefore from now on assume that this holds for for all $i$. Put $y_i = x_i/|\rho_i|$ and $\epsilon_i = sign(\rho_i)$, and define the vector $\boldsymbol{w}^I(\boldsymbol{y})$ by

$$w_i^I(\boldsymbol{y}) = y_i - \epsilon_i \frac{1}{d(I)} \sum_{j\in I} d_j \epsilon_j y_j \quad \text{for } i \in I. \tag{9.28}$$

Then (9.22) takes the form

$$(Q_I^{-1}\boldsymbol{u}_I)_i = \frac{|\rho_i|}{\sigma_i(1-\rho_i^2)} w_i^I(\boldsymbol{y}) \quad \text{for } i \in I. \tag{9.29}$$

The weights thus have the same sign as $w_i^I(\boldsymbol{y})$.

We also have

$$\boldsymbol{u}_I \cdot Q_I^{-1}\boldsymbol{u}_I = L_I(\boldsymbol{y}), \tag{9.30}$$

where

$$L_I(\boldsymbol{y}) = \sum_{i\in I} d_i y_i \big(y_i - \frac{\epsilon_i}{1+d(I)} \sum_{j\in I} d_j \epsilon_j y_j\big) = \tag{9.31}$$

$$\sum_{i\in I} d_i y_i^2 - \frac{1}{1+d(I)} (\sum_{j\in I} d_j \epsilon_j y_j)^2.$$

The next lemma shows that $I_*$ cannot be a proper subset of a set with positive weights.

**Lemma 9.12.** *If* $I \supset I_*$ *and* $w_I(i) > 0$ *for all* $i \in I$*, then* $I = I_*$.

*Proof.* Let $I' = I \setminus I_*$. Then $y_i \leq \epsilon_i \xi_*$ for $i \in I'$, and

$$0 < w_I(i)(1 + d(I)) = y_i(1 + d(I_*)) - \epsilon_i \sum_{j \in I_*} d_j \epsilon_j y_j + \sum_{j \in I'} d_j(y_i - \epsilon_i \epsilon_j y_j).$$

Here

$$y_i(1 + d(I_*)) - \epsilon_i \sum_{j \in I_*} d_j \epsilon_j y_j \leq \epsilon_i \xi_*(1 + d(I_*)) - \epsilon_i \sum_{j \in I_*} d_j \epsilon_j y_j = 0$$

for $i \in I'$. Here we used (9.26). Therefore

$$0 < \sum_{j \in I'} d_j(y_i - \epsilon_i \epsilon_j y_j)$$

for $i \in I'$. Let $i$ be such that $y_i = \min_{j \in I'} y_j$. Then

$$y_i - \epsilon_i \epsilon_j y_j = y_i - y_j \leq 0 \text{ if } \epsilon_j = \epsilon_i$$

and

$$y_i - \epsilon_i \epsilon_j y_j = y_i + y_j \leq \epsilon_i \xi_* + \epsilon_j \xi_* = 0 \text{ if } \epsilon_j = -\epsilon_i.$$

Therefore $0 < 0$ if $I' \neq \emptyset$. □

We shall also use the following inequality.

**Lemma 9.13.** *If* $w_i^I(\boldsymbol{y}) > 0$ *for* $i \in I$ *and* $\boldsymbol{y}' \geq \boldsymbol{y}$*, then*

$$L_I(\boldsymbol{y}') \geq L_I(\boldsymbol{y}) \tag{9.32}$$

*with equality only if* $y_i' = y_i$ *for* $i \in I$.

*Proof.*

$$L_I(\boldsymbol{y}') = L_I(\boldsymbol{y}) + (\boldsymbol{y}' - \boldsymbol{y}) \cdot gradL_I(\boldsymbol{y}) + \frac{1}{2}(\boldsymbol{y}' - \boldsymbol{y}) \cdot L_I''(\boldsymbol{y}' - \boldsymbol{y}). \tag{9.33}$$

Here

$$\frac{\partial L_I(\boldsymbol{y})}{\partial y_i} = 2 d_i w_i^I(\boldsymbol{y}) > 0,$$

and $L_I''$ is the positive definite matrix with elements

$$\frac{\partial^2 L_I(\boldsymbol{y})}{\partial y_i \partial y_j} = 2(d_i \delta_{i,j} - \frac{d_i \epsilon_i d_j \epsilon_j}{1 + d(I)}).$$

□

*Proof of the theorem.* Define

$$I_1 = I \cap I_*,\ I_2 = I \setminus I_1,\ I_0 = I_* \setminus I_1,$$

and put for $j = 0, 1, 2$

$$D_j = \sum_{i \in I_j} d_j,\ S_j = \sum_{i \in I_j} d_j \epsilon_j y_j.$$

If $I_0 = \emptyset$, then by Lemma 9.12 we have $I = I_*$.

Assume that $I_0 \neq \emptyset$.

We have

$$L_{I_*}(\boldsymbol{y}) = \sum_{i \in I_0} d_i y_i^2 + \sum_{i \in I_1} d_i y_i^2 - \frac{(S_0 + S_1)^2}{1 + D_0 + D_1}.$$

Therefore by Schwarz inequality

$$L_{I_*}(\boldsymbol{y}) \geq \frac{S_0^2}{D_0} + \sum_{i \in I_1} d_i y_i^2 - \frac{(S_0 + S_1)^2}{1 + D_0 + D_1} \tag{9.34}$$

with equality only if $\epsilon_i y_i = c$ for all $i \in I_0$ and some constant $c$.

Let $y_i' = \epsilon_i \xi_*$ if $i \in I_2$ and $y_i' = y_i$ otherwise. Then $y_i' \geq y_i$ for $i \in I$, and hence by Lemma 9.13

$$L_I(\boldsymbol{y}) \leq \sum_{i \in I_1} d_i y_i^2 + D_2 \xi_*^2 - \frac{(S_1 + D_2 \xi_*)^2}{1 + D_1 + D_2}. \tag{9.35}$$

It follows from (9.34) and (9.35) that

$$L_{I_*}(\boldsymbol{y}) - L_I(\boldsymbol{y}) \geq \frac{S_0^2}{D_0} - \frac{(S_0 + S_1)^2}{1 + D_0 + D_1} - D_2 \xi_*^2 + \frac{(S_1 + D_2 \xi_*)^2}{1 + D_1 + D_2}. \tag{9.36}$$

Put

$$\eta = \xi_* - \frac{S_1}{1 + D_1}.$$

Then

$$S_0 + S_1 = (1 + D_0 + D_1)\xi_*,\ S_0 = D_0 \xi_* + (1 + D_1)\eta,\ S_1 = (1 + D_1)(\xi_* - \eta),$$

and the expression to the right in (9.36) simplifies to

$$\frac{(1 + D_1)^2 (1 + D_0 + D_1 + D_2)}{D_0 (1 + D_1 + D_2)} \eta^2.$$

We have thus shown that

$$L_{I_*}(\boldsymbol{y}) \geq L_I(\boldsymbol{y}). \tag{9.37}$$

Assume that equality holds. Then $\eta = 0$, that is,

$$\frac{S_1}{1+D_1} = \xi_*, \tag{9.38}$$

and we must have equality also in (9.34), which implies $S_0 = cD_0$. Therefore by (9.26)

$$\xi_* = \frac{S_0 + S_1}{1 + D_0 + D_1} = \frac{cD_0 + \xi_*(1 + D_1)}{1 + D_0 + D_1},$$

and hence $c = \xi_*$, since $D_0 > 0$. This implies $y_i = \epsilon_i \xi_*$ for $i \in I_0$, contradicting (9.25). Therefore equality holds in (9.37) only if $I = I_*$. □

*Proof of the corollary.* We shall show that if $I \supset I_*$, then $J \supset I_*$. The corollary then follows from Lemma 9.12.

First note that $J$ will always contain $\{i \in [1:m]; x_i > 0, \rho_i = 0\}$. Then assume that $\rho_i \neq 0$ for all $i$ and switch to $y_i$.

Let $i \in I_*$. Then

$$y_i(1 + d(I)) - \sum_{j \in I} d_j y_j = y_i(1 + d(I_*)) - \sum_{j \in I_*} d_j y_j + d(I \setminus I_*) y_i - \sum_{j \in I \setminus I_*} d_j y_j >$$

$$d(I \setminus I_*) y_i - \sum_{j \in I \setminus I_*} d_j y_j = \sum_{j \in I \setminus I_*} d_j (y_i - y_j) \geq \sum_{j \in I \setminus I_*} d_j (\xi_* - \xi_*) = 0.$$

Here we used that $y_i > \xi_*$ since $i \in I_*$, and that $y_j \leq \xi_*$ since $j \notin I_*$. □

It remains to show that the optimal set (9.13) coincides with (9.25).

**Lemma 9.14.** *Assume that $\rho \geq 0$ and that $\rho_i = \sqrt{\rho}$ for $i = 1, ..., m$.. The sets (9.13) and (9.25) coincide.*

*Proof.* Let $f(\xi)$ be as in Lemma 9.7 and $I_*$ as in (9.25). Put

$$g(\eta) = (1 - \rho)\eta + \rho \sum_{i=1}^{m} \min(0, \eta - x_i).$$

Then

$$f(\xi) = \frac{g(\xi\sqrt{\rho})}{\sqrt{\rho}(1 - \rho)}$$

and hence

$$I_* = \{i \in [1:m]; x_i > \eta_*\},$$

where $g(\eta_*) = 0$. □

## 9.3 NOTES

My source here has been Elton and Gruber (1981).

# APPENDIX A

# ANSWERS AND SOLUTIONS TO EXERCISES

## Chapter 1

**1.1.** Let $R'$ and $R''$ stand for the growth factors for the two parts of the period. Then $R' = 1 + r'$ and $R'' = 1 + r''$, and hence the growth factor for the entire period equals $R'R'' = (1 + r')(1 + r'') = 1 + r' + r'' + r'r''$.

**1.2.** In this case $R' = e^{r'}$ and $R'' = e^{r''}$, and hence $R'R'' = e^{r'}e^{r''} = e^{r'+r''}$.

**1.3.** $PV(417) = 417/1.05 = 397.14$, $PV(430) = 430/1.05^2 = 390.02$.

**1.4.** $T = (\ln 2)/r = 13.86$ years $=$ 13 years 10 months and 11 days.

*Mathematical Asset Management.* By Thomas Höglund

**1.5.** 15443 EUR.

**1.6.** After six years one has 1.34, 1.37, and 1.30, respectively. Harvesting after 2 years is thus preferable.

**1.7.** Define $P(d) = x_0 + x_1 d + \cdots + x_n d^n$. Then $P(d) = |x_0| - (|x_1|d + \cdots + |x_n|d^n)$. It follows that $P$ is a decreasing function for $d \geq 0$ and that $P(0) > 0$, $P(1) < 0$. Therefore $P$ has a unique zero, $d_0$, satisfying $0 < d_0 < 1$.

The second case follows in a similar way from the identity $c^n P(d) = |x_n| - (|x_{n-1}|c + \cdots + |x_0|c^n)$. Here $c = 1/d$.

**1.8.** Continuous rate $= 0.111$, return rate $= 0.118$ per year.

**1.9.** The effective rates are given by the yearly returns 0.050, 0.054, and 0.045 (i.e., the continuous rates 0.049, 0.053, and 0.044), respectively.

**1.10.** The loan and the saving have the present values

$$S - K(d + \cdots + d^n) \text{ and } -K((1 + d + \cdots + d^{n-1}) + Sd^n,$$

respectively. The statement now follows from the identity

$$1 + d + \cdots + d^{n-1} = \frac{1 - d^n}{1 - d}.$$

**1.11.** The tangent line to $F$ in the point $x_{k-1}$ has the equation

$$y - F(x_{k-1}) = F'(x_{k-1})(x - x_{k-1}),$$

and this line cuts the $x$-axis ($y = 0$) in the point $x = x_k$.

**1.12.** Continuous rate $= 0.084$, return $= 0.087$ per year.

**1.13.** Yearly return $= 6.2\%$, 81.30 months, 386.7 EUR.

**1.14.** (a)

$$P = \frac{c}{m} \sum_{k=1}^{n} d^k + d^n F = \frac{c}{m} d \frac{1 - d^n}{1 - d} + d^n F.$$

(b)

$$P = \frac{c}{m} \frac{d}{1 - d} + d^n (F - \frac{c}{m} \frac{d}{1 - d}) = \frac{c}{y} + d^n (F - \frac{c}{y}).$$

**1.15.** It follows from Exercise 1.14b that

$$P_1 - P_2 = (d^{n_1} - d^{n_2})(F - \frac{c}{y}).$$

The first factor here is positive since $n_1 < n_2$, and hence $P_1 > P_2$ if and only if $F > \frac{c}{y}$.

**1.16.** Yearly return $= 4.1\%$.

**1.17.** (a) $P = \frac{c_2 - c}{c_2 - c_1} P_1 + \frac{c - c_1}{c_2 - c_1} P_2$; (b) $\frac{c_2}{c_2 - c_1} P_1/P$ and $-\frac{c_1}{c_2 - c_1} P_2/P$; (c) For $c$ between $c_1$ and $c_2$.

**1.18.** $3r_l > r_b/(1 + r_b)$, where $r_l$ stands for the yearly lending rate of return and $r_b$ for the interest on deposit.

**1.19.** (a) The effective rate of return is 0 or 100% for both.

(b) At $t = 0$: Borrow 1000 EUR and accept the second cash stream.

At $t = 1$: Pay off the loan with $1000 \times 1.05 = 1050$ EUR. Lend the remainder $3000 - 1050 = 1950$ EUR.

At $t = 2$: The loan is repayed to you with $1950 \times 1.04 = 2028$ EUR and you pays 2000 EUR. Left 28 EUR. You thus get the cash stream $(0, 0, 28)$.

**1.20.** (a) $r_1(\boldsymbol{x}) = 4\%$, $r_1(\boldsymbol{y}) = 5\%$ ; (b) $PV(\boldsymbol{x}) = -1.11$, $PV(\boldsymbol{y}) = 2.94.$, $r_1 = 3\%$, $r_2 = 6\%$ (rounded); (c) Accept the loan and buy 426 one-year zero-coupon bonds and 656 two year zero-coupon bonds. This gives you the cash stream (2.94, 0, 0); (d) The lender short-sells 866 one year zero-coupons and 181 two year zero-coupons. This gives the lender the cash stream $(1.11, 0, 0)$.

**1.21.** (a) $L_2$; (b) $p = 2$.

**1.22.** Complete for all $p$. No arbitrage for $1 < p < 2$.

**1.23.** (a) Price: 98.19. Effective rate: 10.74%; (b) Prices: 101.38, 101.76, 96.37. Effective rates: 8.50%, 9.02%, 11.79%.

**1.24.** Let $\xi_0, ..., \xi_{n-1}$ be real numbers such that

$$\xi_0 \boldsymbol{r}^0 + \cdots + \xi_{n-1} \boldsymbol{r}^{n-1} = 0.$$

The polynomial $P(r) = \xi_0 + \xi_1 r + \cdots + \xi_{n-1} r^{n-1}$ then has the $n$ different zeros, $r_1, ..., r_n$, and it is of degree $n - 1$. Therefore it is identically zero, and hence $\xi_0 = \cdots = \xi_{n-1} = 0$. It follows that $\boldsymbol{r}^0, ..., \boldsymbol{r}^{n-1}$ are $n$ linearly independent vectors in $\mathbb{R}^n$.

**1.25.** We have $\frac{\partial P}{\partial r} = -Te^{-rT} = -TP$, and it follows from Taylors formula that $P + \partial P = P + \frac{\partial P}{\partial r}\partial r + O((\partial r)^2)$.

**1.26.**

$$\frac{\partial P}{\partial r} = \frac{\partial e^{-kr_k}}{\partial r_k} x_k = -ke^{-kr_k} x_k = -kd_k x_k.$$

**1.27.** The portfolio consisting of the two bonds has the duration

$$\frac{\frac{\partial(P_1+P_2)}{\partial r}}{P_1 + P_2}.$$

We have $\frac{\partial P_i}{\partial r} = P_i D_i$, and hence $\frac{\partial(P_1+P_2)}{\partial r} = P_1 D_1 + P_2 D_2$ from which the result follows.

**1.28.** Prices: $98.67, 95.45, 90.70$. Durations: $4.54, 3.77, 2.00$.

**1.29.** (a) The price $P$ is as in Exercise 1.14, and hence

$$\frac{\partial P}{\partial r} = \frac{c}{m}\sum_{k=1}^{n} kd^{k-1}\frac{\partial d}{\partial r} + nd^{n-1}\frac{\partial d}{\partial r}F.$$

But $\frac{\partial d}{\partial r} = -\frac{1}{m}d$, and hence

$$\frac{\partial P}{\partial r} = -\frac{c}{m^2}\sum_{k=1}^{n} kd^k - \frac{n}{m}d^n F = -\frac{1}{m}(\frac{c}{m}\sum_{k=1}^{n} kd^k - nd^n F).$$

(b)

$$\sum_{k=1}^{n} kd^{k-1} = \frac{\partial}{\partial d}\sum_{k=0}^{n} d^k = \frac{\partial}{\partial d}\frac{1-d^{n+1}}{1-d} = \frac{-(n+1)d^n}{1-d} + \frac{1-d^{n+1}}{(1-d)^2}.$$

(c) It follows from (a) and (b) that

$$PD = \frac{1}{m}(\frac{c}{m}\frac{d}{1-d}(-(n+1)d^n + \frac{1-d^{n+1}}{1-d}) + nd^n F) =$$

$$\frac{1}{m}(\frac{c}{m}\frac{d}{1-d}(-nd^n + \frac{1-d^n}{1-d}) + nd^n F).$$

We have

$$\frac{d}{1-d} = \frac{m}{y} \text{ and } \frac{1}{1-d} = \frac{y+m}{y},$$

and hence

$$PD = \frac{1}{m}(\frac{c}{y}(-nd^n + \frac{y+m}{y}(1-d^n)) + nd^nF).$$

The result follows from this and the identity $n = mT$.

(d) This follows from (c) and Exercise 1.14b.

**1.30.** $D^{(1)}$: $0.235, 0.194, 0.098$. $D^{(2)}$: $0.012, 0.010, 0.005$.

**1.31.** Buy A for 150.83 EUR and C for 279.64 EUR.

**1.32.** (a) $T = D$; (b) $Tr_T = D^{(1)}$.

**1.33.** (a) $v_{T_1} = \frac{T_2-T}{T_2-T_1}$, $v_{T_2} = \frac{T-T_1}{T_2-T_1}$.

(b) Let $r_1, r, r_2$ stand for the interest rates at time 0 with durations $T_1, T, T_2$, respectively. Also let $a_{T_i}$ denote the number of zero-coupon bonds with duration $T_i$, $i = 1, 2$. Then $a_{T_i}e^{-r_iT_i} = d_Tv_{T_i}$, where $d_T = e^{-rT}$.

We have

$$\Delta_0(\boldsymbol{r} + \mathbf{1}\partial p) = a_1e^{-(r_1+\partial p)T_1} + a_2e^{-(r_2+\partial p)T_2} - e^{-(r+\partial p)T} =$$

$$d_T(v_{T_1}e^{-\partial pT_1} + v_{T_2}e^{-\partial pT_2} - e^{-\partial pT}).$$

Using the expansion

$$e^{-x} = 1 - x + \frac{x^2}{2} + O(|x|^3),$$

we get

$$\Delta_0(\boldsymbol{r} + \mathbf{1}\partial p) =$$

$$d_T(v_{T_1} + v_{T_2} - 1 - (v_{T_1}T_1 + v_{T_2}T_2 - T)\partial p + (v_{T_1}T_1^2 + v_{T_2}T_2^2 - T^2)\frac{(\partial p)^2}{2} +$$

$$O(|\partial p|^3)) =$$

$$d_T(0 - 0\cdot\partial p + (-T^2 + T(T_1+T_2) - T_1T_2))\frac{(\partial p)^2}{2} + O(|\partial p|^3)) =$$

$$d_T(T-T_1)(T_2-T)\frac{(\partial p)^2}{2} + O(|\partial p|^3).$$

Here we used the expressions for the weights given in (a).

## Chapter 2

**2.1.** We have $\ln P = \ln E - \ln r$, and

$$\partial \ln E = \ln(E + \partial E) - \ln E = \ln(1 + \frac{\partial E}{E}) = \frac{\partial E}{E} + O((\partial E)^2),$$

and similar for $\partial \ln r$.

**2.2.** $E/r$.

**2.4.** We have $\partial d_k = -kd_k \partial r_k$, and hence

$$\partial P = \sum d_k \partial E_k + \sum E_K \partial d_k == \sum d_k \partial E_k - \sum E_K k d_k \partial r_k,$$

from which the identity follows.

**2.5.** We have $(\frac{1}{1+r_{max}})^k \leq d_k \leq (\frac{1}{1+r_{min}})^k$. Also

$$\sum_{k=1}^{\infty} (\frac{1}{1+r})^k = \frac{1}{1+r} \sum_{k=0}^{\infty} (\frac{1}{1+r})^k = \frac{1}{1+r} \frac{1}{1 - \frac{1}{1+r}} = \frac{1}{r}.$$

Using this with $r = r_{min}$ and $r = r_{max}$ gives the inequalities.

**2.6.** Proof of (b): Let $K_1 < K_2$. We shall show that

$$C_0(K_1) - C_0(K_2) - e^{-r(T)}(K_2 - K_1) \leq 0.$$

Assume that this inequality does not hold. Buy a call with strike price $K_2$, write a call with strike price $K_1$, and put the income

$$I = C_0(K_1) - C_0(K_2)$$

in the cash account. Wait until time $T$. The cash has now increased to $e^{rT} I > K_2 - K_1$, and the value of the options portfolio is now $C_T(K_2) - C_T(K_1) = \max(0, S_T - K_2) - \max(0, S_T - K_1)$. The value of the portfolio therefore equals

$$Ie^{rT} + \max(0, S_T - K_2) - \max(0, S_T - K_1) >$$

$$K_2 - K_1 + \max(0, S_T - K_2) - \max(0, S_T - K_1) =$$

$$\max(K_2, S_T) - \max(K_1, S_T) \geq 0.$$

**2.7.** Buy a call, and write a put with strike prices equal to the delivery price and with the same maturity date as the forward.

**2.8.** The difference from a stock that pays no dividends is that one at time zero borrows (or lends) the additional amount $De^{-rt}$.

**2.11.** (a) Buy one call with strike price $K_1$ and one with strike price $K_3$. Write two calls with strike price $K_2$.

(b) Buy one call and one put both with strike price $K$.

(c) Buy a call with strike price $K_2$ and a put with strike price $K_1$.

**2.12.** (a) $S_T \geq e^{rT}S_0(1+p)$. (b) $S_T \leq K/(1+p)$. (c) No value.

**2.13.** (a) Keep $e^{-rT}K$ EU in cash and buy a call with the same expiration date and strike price as the put. The value of the portfolio at $t$ equals $e^{-r(T-t)}K + C_t(K)$ which according to the put-call parity $= F_t$.

(b) In the event of a crash on the stock market, it may happen that it is not possible to exercise the put because the collateral of the writer has decreased in value. This is an advantage of the cash-call portfolio even if there are similar but less probable objections against this.

**2.14.** (a) $P_0(K) = C_0(K) - (S_0 - e^{-rT}K)$, (b) Otherwise $C_0(H) \geq C_0(e^{rT}S_0) = P_0(e^{rT}S_0) > P_0(K)$.

**2.16.** (a) $S_T \geq S_0e^{rT}$. (b) $S_T \leq H$. (c) $S_0e^{rT} \leq S_T \leq H$.

**2.17.** (a) Keep $e^{-rT}K$ EUR in cash. Buy a call with strike price $K$ and write a call with strike price $H$.

**2.18.** (a) $S_T \geq (1-c)S_0e^{rT}$. (b) $S_T \leq H/(1-c)$. (c) $(1-c)S_0e^{rT} \leq S_T \leq H/(1-c)$.

**2.19.** (a) Keep $e^{-rT}K$ in cash and write a put with strike price $H$.

**2.20.** (b)

| $S_T$ | Strategy no. |
|---|---|
| $0 < S_T < (1-c)K$ | 2 |
| $(1-c)K < S_T < \frac{1+p}{1-c}H$ | 3 |
| $\frac{1+p}{1-c}H < S_T < \infty$ | 1 |

(c)

| $S_T$ | Strategy no. |
|---|---|
| $0 < S_T < \frac{1-c}{1+p}K$ | 3 |
| $\frac{1-c}{1+p}K < S_T < (1+p)H$ | 1 |
| $(1+p)H < S_T < \infty$ | 2 |

**2.21.** (a) 1762.79 USD. (b) 13.09 USD.

**2.22.** 9.69 EUR.

**2.23.** (a) 11.00 EUR. (b) Cash streams $(-2.51, \max(0, S_2 - 11.00)), (0, S_2 - 11.00)$. The first has higher present value if $S_2 < 8.24$.

## Chapter 3

**3.2.**

$$f = c + as = f_u - as_u + as = f_u - \frac{f_u - f_d}{s_u - s_d}(s_u - s) = f_u q + f_d(1 - q).$$

**3.3.**

$$P: \begin{array}{ccc} & & 19 \\ & 19 & \\ 20 & & 19 \\ & 21 & \\ & & 23 \end{array} \quad a: \begin{array}{ccc} & & \cdot \\ & 0 & \\ -0.5 & & \cdot \\ & -1 & \\ & & \cdot \end{array} \quad c \ \begin{array}{ccc} & & \cdot \\ & 19 & \\ 30 & & \cdot \\ & 39 & \\ & & \cdot \end{array}$$

**3.4.**

$$P: \begin{array}{ccc} & & 80 \\ & 40 & \\ 20 & & 0 \\ & 0 & \\ & & 0 \end{array} \quad a: \begin{array}{ccc} & & \cdot \\ & 20 & \\ 10 & & \cdot \\ & 0 & \\ & & \cdot \end{array} \quad c: \begin{array}{ccc} & & \cdot \\ & -400 & \\ -180 & & \cdot \\ & 0 & \\ & & \cdot \end{array}$$

**3.5.** (a)

$$P: \begin{array}{ccccc} & & & & 2 \\ & & & 1 & \\ & & 1/2 & & 0 \\ & 1/4 & & 0 & \\ 1/8 & & 0 & & 0 \\ & 0 & & 0 & \\ & & 0 & & 0 \\ & & & 0 & \\ & & & & 0 \end{array} \quad a: \begin{array}{ccccc} & & & & \cdot \\ & & & 1 & \\ & & 1/2 & & \cdot \\ & 1/4 & & 0 & \\ 1/8 & & 0 & & \cdot \\ & 0 & & 0 & \\ & & 0 & & \cdot \\ & & & 0 & \\ & & & & \cdot \end{array} \quad c: \begin{array}{ccccc} & & & & \cdot \\ & & & -12 & \\ & & -5.5 & & \cdot \\ & -2.5 & & 0 & \\ -1.125 & & 0 & & \cdot \\ & 0 & & 0 & \\ & & 0 & & \cdot \\ & & & 0 & \\ & & & & \cdot \end{array}$$

**3.5.** (b)

$$
P:\ \begin{array}{ccccc}
 & & & & 0\\
 & & & 0 & \\
 & & 0 & & 0\\
 & 0 & & 0 & \\
1/8 & & 0 & & 0\\
 & 1/4 & & 0 & \\
 & & 1/2 & & 0\\
 & & & 1 & \\
 & & & & 2
\end{array}
\qquad
a:\ \begin{array}{ccccc}
 & & & & \cdot\\
 & & & 0 & \\
 & & 0 & & \cdot\\
 & 0 & & 0 & \\
-1/8 & & 0 & & \cdot\\
 & -1/4 & & 0 & \\
 & & -1/2 & & \cdot\\
 & & & -1 & \\
 & & & & \cdot
\end{array}
\qquad
c:\ \begin{array}{ccccc}
 & & & & \cdot\\
 & & & 0 & \\
 & & 0 & & \cdot\\
 & 0 & & 0 & \\
1.375 & & 0 & & \cdot\\
 & 2.5 & & 0 & \\
 & & 4.5 & & \cdot\\
 & & & 8 & \\
 & & & & \cdot
\end{array}
$$

**3.5.** (c)

$$
P:\ \begin{array}{ccccc}
 & & & & 12\\
 & & & 12 & \\
 & & 11.5 & & 12\\
 & 10.75 & & 11 & \\
10 & & 10 & & 10\\
 & 9.25 & & 9 & \\
 & & 8.5 & & 8\\
 & & & 8 & \\
 & & & & 8
\end{array}
\qquad
a:\ \begin{array}{ccccc}
 & & & & \cdot\\
 & & & 0 & \\
 & & 1/2 & & \cdot\\
 & 3/4 & & 1 & \\
3/4 & & 1 & & \cdot\\
 & 3/4 & & 1 & \\
 & & 1/2 & & \cdot\\
 & & & 0 & \\
 & & & & \cdot
\end{array}
\qquad
c:\ \begin{array}{ccccc}
 & & & & \cdot\\
 & & & 12 & \\
 & & 5.5 & & \cdot\\
 & 2.5 & & 0 & \\
2.5 & & 0 & & \cdot\\
 & 2.5 & & 0 & \\
 & & 4.5 & & \cdot\\
 & & & 8 & \\
 & & & & \cdot
\end{array}
$$

**3.6.** The proof is by induction. The identity holds for $k = 0$. Assume that it holds for $k$. We shall show that it holds for $k + 1$.

$$F_{n-k-1}(s) = \frac{1}{2}[F_{n-k}(s+h) + F_{n-k}(s-h)] =$$

$$\frac{1}{2^{k+1}}[\sum_{j=0}^{k} \binom{k}{j} f(s+h+(k-2j)h) + \sum_{j=0}^{k} \binom{k}{j} f(s-h+(k-2j)h)] =$$

$$\frac{1}{2^{k+1}}[\sum_{j=0}^{k} \binom{k}{j} f(s+(k+1-2j)h) + \sum_{j=1}^{k+1} \binom{k}{j-1} f(s+(k+1-2j)h)] =$$

$$\frac{1}{2^{k+1}} \sum_{j=0}^{k+1} \binom{k+1}{j} f(s+(k+1-2j)h).$$

Here we used Pascal's triangle:

$$\binom{k}{j-1} + \binom{k}{j} = \binom{k+1}{j}.$$

Note that this also holds for $j = 0$ and $j = k+1$ if we interpret $\binom{k}{-1} = 0$ and $\binom{k}{k+1} = 0$.

**3.7.**

$$
P:\ \begin{array}{ccc} & & 22.4 \\ & 21.4 & \\ & & 20.4 \\ 20 & & \\ & & 18.6 \\ & 18.6 & \\ & & 18.6 \end{array}
\qquad
a:\ \begin{array}{ccc} & & \cdot \\ & 0.47 & \\ & & \cdot \\ 0.70 & & \\ & & \cdot \\ & 0 & \\ & & \cdot \end{array}
\qquad
c:\ \begin{array}{ccc} & & \cdot \\ & 11.0 & \\ & & \cdot \\ 6.0 & & \\ & & \cdot \\ & 18.6 & \\ & & \cdot \end{array}
$$

**3.8.**

$$
P:\ \begin{array}{ccc} & & 22.0 \\ & 21.4 & \\ & & 20.6 \\ 20 & & \\ & & 19.4 \\ & 18.6 & \\ & & 18.0 \end{array}
\qquad
a:\ \begin{array}{ccc} & & \cdot \\ & 0.33 & \\ & & \cdot \\ 0.66 & & \\ & & \cdot \\ & 0.33 & \\ & & \cdot \end{array}
\qquad
c:\ \begin{array}{ccc} & & \cdot \\ & 14.0 & \\ & & \cdot \\ 6.6 & & \\ & & \cdot \\ & 12.6 & \\ & & \cdot \end{array}
$$

**3.9.** (a) $\frac{1-d}{u-d}$; (b) $\frac{1}{1+u}$; (c) $\frac{1}{2}$.

**3.10.**

$$
P:\ \begin{array}{cccc} & & & 14.6 \\ & & 8.8 & \\ & 5.0 & & 4 \\ 2.7 & & 1.8 & \\ & 0.8 & & 0 \\ & & 0 & \\ & & & 0 \end{array}
\qquad
a:\ \begin{array}{cccc} & & & \cdot \\ & & 1 & \\ & 0.79 & & \cdot \\ 0.57 & & 0.55 & \\ & 0.30 & & \cdot \\ & & 0 & \\ & & & \cdot \end{array}
$$

**3.11.** This is similar to Exercise 3.6.

$$F_{n-k-1}(s) = qF_{n-k}(su) + (1-q)F_{n-k}(sd) =$$

$$q\sum_{j=0}^{k} \binom{k}{j} q^j (1-q)^{k-j} f(suu^j d^{k-j}) +$$

$$(1-q)\sum_{j=0}^{k} \binom{k}{j} q^j (1-q)^{k-j} f(sdu^j d^{k-j}) =$$

$$\sum_{j=1}^{k+1} \binom{k}{j-1} q^j (1-q)^{k+1-j} f(su^j d^{k+1-j}) +$$

$$\sum_{j=0}^{k} \binom{k}{j} q^j (1-q)^{k+1-j} f(su^j d^{k+1-j}) =$$

$$\sum_{j=0}^{k} \binom{k+1}{j} q^j (1-q)^{k+1-j} f(su^j d^{k+1-j}).$$

**3.12.**

$$\begin{array}{lcccc}
 & & & & 5.63 \\
 & & & 4.40 & \\
 & & 3.53 & & 3.16 \\
P: & 2.91 & & 2.66 & \\
 & & 2.29 & & 2.16 \\
 & & & 1.92 & \\
 & & & & 1.67
\end{array}
\qquad
\begin{array}{lcccc}
 & & & & 1.93 \\
 & & & 1.51 & \\
 & & 1.21 & & 1.09 \\
P_{\text{norm}}: & 1 & & 0.91 & \\
 & & 0.79 & & 0.74 \\
 & & & 0.66 & \\
 & & & & 0.57
\end{array}$$

**3.13.**

$$q = \frac{1}{2}\frac{1}{1+t} = \frac{1}{2}(1 - t + t^2 + O(t^3)),$$

where

$$t = \frac{e^{\delta} - 1}{2} = \frac{\delta}{2} + \frac{\delta^2}{4} + O(\delta^3).$$

This gives the first identity.

We have

$$E[Y_k] = 2\delta(E[X_k] - \frac{k}{2}), \ \text{Var}(Y_k) = (2\delta)^2 \text{Var}(X_k),$$

and

$$E[X_k] = kq, \ \text{Var}(X_k) = kq(1-q)$$

since $X_k$ is binomially distributed $(k, q)$. The second and third identity follow from this and the first identity.

## Chapter 4

**4.1.**

$$G(t_1, t_2) = \ln \frac{S_{t_2}}{S_{t_1}} = \ln \frac{S_{t_3}}{S_{t_1}} \frac{S_{t_2}}{S_{t_3}} = \ln \frac{S_{t_3}}{S_{t_1}} + \ln \frac{S_{t_2}}{S_{t_3}} = G(t_1, t_3) + G(t_3, t_2).$$

**4.2.** (a) We have $G(0, t+s) = G(0, t) + G(t, t+s)$, and hence

$$\nu(t+s) = EG(0,t) + EG(t, t+s) = EG(0,t) + EG(0,s) = \nu(t) + \nu(s).$$

Here we used part (b) of the moment assumption. Also

$$\sigma^2(t+s) = \text{Var}(G(0,t) + G(t,t+s)) =$$

$$\text{Var}(G(0,t)) + \text{Var}(G(t,t+s)) + \text{Cov}(G(0,t), G(t,t+s)) =$$

$$\text{Var}(G(0,t)) + \text{Var}(G(0,s)) = \sigma^2(t) + \sigma^2(s).$$

Here we used part (c) and (b) of the moment assumption.

(b) Let $m$ and $n$ be positive integers, then

$$\nu(\frac{m}{n}) = \nu(\frac{1}{n}) + \nu(\frac{m-1}{n}) = \cdots = m\nu(\frac{1}{n}),$$

and

$$\nu(1) = \nu(\frac{n}{n}) = n\nu(\frac{1}{n}).$$

It follows that $\nu(\frac{m}{n}) = \nu(1)\frac{m}{n}$, that is, $\nu(t) = \nu(1)t$ for all rational $t > 0$. This holds therefore for all real $t \geq 0$ since $\nu$ is continuous.

The proof for $\sigma^2(t)$ is the same.

**4.3.** By Exercise 4.1

$$\sum_{k=1}^{n} x_k = \sum_{k=1}^{n} G((k-1)\partial t, k\partial t) = G(0,T).$$

Therefore

$$\hat{\nu} = \frac{G(0,T)}{n\partial t} = \frac{\ln(S_T/S_0)}{T}.$$

**4.4.** (a) $(-2\%, 92\%)$. (b) $T \approx 1500$ years.

**4.5.** (a) Let

$$X_n = n\hat{\sigma}^2/\sigma^2.$$

Then $X_n$ is $\chi^2$ distributed with $n-1$ degrees of freedom and hence has the expectation $n-1$ and variance $2(n-1)$. The identities now follow from the identity $\hat{\sigma}^2 = \frac{\sigma^2}{n}X_n$.

(b) The $\chi^2$ variable $X_n$ is the sum of the squares of $n-1$ independent standardized normal variables, and hence by the central limit theorem $X_n$ is asymptotically normally distributed with expectation $n-1$ and variance $2(n-1)$. Let $Z_n = \frac{X_n - n}{\sqrt{2n}}$. Then

$$Z_n = \frac{X_n - (n-1)}{\sqrt{2(n-1)}}\sqrt{1 - \frac{1}{n}} - \frac{1}{\sqrt{2n}}$$

and hence $Z_n$ is asymptotically normally distributed with expectation 0 and variance 1. We have

$$\hat{\sigma} = \sqrt{\hat{\sigma}^2} = \sqrt{\frac{\sigma^2}{n}(n + \sqrt{2n}Z_n)} = \sigma(1 + \sqrt{\frac{2}{n}}Z_n)^{\frac{1}{2}} = \sigma(1 + \frac{1}{\sqrt{2n}}Z_n + O(1/n)).$$

It follows that $\hat{\sigma}$ is asymptotically normally distributed with expectation $\sigma$ and variance $\frac{\sigma^2}{2n}$. Therefore

$$\frac{\hat{\sigma}}{1 \pm \frac{z_{\epsilon/2}}{\sqrt{2n}}}$$

is a confidence interval with the approximate confidence level $1 - \epsilon$. But

$$\frac{\hat{\sigma}}{1 \pm \frac{z_{\epsilon/2}}{\sqrt{2n}}} = \hat{\sigma}(1 \mp \frac{z_{\epsilon/2}}{\sqrt{2n}} + O(1/n)),$$

and hence also the interval

$$\hat{\sigma}(1 \mp \frac{z_{\epsilon/2}}{\sqrt{2n}})$$

has the approximate confidence level $1 - \epsilon$.

**4.6.**

$$\text{Var}(\breve{\sigma}^2) = \frac{1}{(n\partial t)^2} \sum_{i,j} \text{Cov}(X_i^2, X_j^2) =$$

$$\frac{1}{(n\partial t)^2}(\sum_i \text{Var}(X_i^2) + 2\sum_{i<j} \text{Cov}(X_i^2, X_j^2)).$$

We have

$$\text{Var}(X_i^2) = \text{Var}(X_1^2) \text{ and } \text{Cov}(X_i^2, X_j^2) = \text{Cov}(X_1^2, X_{j-i+1}^2) = \text{Var}(X_1^2)\rho_{j-i}$$

for $i < j$. Therefore

$$\text{Var}(\breve{\sigma}^2) = \frac{1}{(n\partial t)^2}\text{Var}(X_1^2)(n + 2\sum_{i<j} \rho_{j-i}).$$

The identity now follows from

$$\sum_{i<j} \rho_{j-i} = \sum_k (n-k)\rho_k.$$

**4.7.** (a)

$$\breve{\sigma} = \sqrt{\breve{\sigma}^2} = \sqrt{\sigma^2(1 + 2Zd)} = \sigma(1 + Zd + O(d^2)).$$

(b) It is seen in the same way as in Exercise 4.5b that

$$\breve{\sigma}(1 \pm z_{\epsilon/2}d)$$

is a confidence interval with the approximate confidence level $1 - \epsilon$. The confidence level will be approximately the same if we replace $d$ by the estimated value $\hat{d}$.

**4.8.**

$$d = \sqrt{\frac{\text{Var}(\breve{\sigma}^2)}{2\sigma^2}} = \sqrt{\frac{\partial t}{2T}}$$

**4.9.** (a) 15, 5, 2 years, respectively. (b) 4, 1, 1/2 years, respectively.

**4.10.**

| | AFGX | AZN | LME | HM | SDIA | SKA |
|---|---|---|---|---|---|---|
| 90% | (0.19, 0.27) | (0.28, 0.35) | (0.42, 0.56) | (0.34, 0.52) | (0.39, 0.57) | (0.23, 0.31) |
| 50% | (0.22, 0.25) | (0.30, 0.34) | (0.46, 0.52) | (0.39, 0.47) | (0.44, 0.52) | (0.25, 0.29) |

**4.11.**

$$1 + R(t_1, t_2) = \frac{S(t_2)}{S(t_1)} = \frac{S(t_3)}{S(t_1)}\frac{S(t_2)}{S(t_3)} = (1 + R(t_1, t_3))(1 + R(t_3, t_2)).$$

**4.12.**

$$ER(0, \partial t) = Ee^{G(0,\partial t)} - 1 = e^{\nu\partial t}Ee^{\sigma\sqrt{\partial t}Z} - 1 =$$

$$e^{\nu\partial t}e^{\frac{\sigma^2\partial t}{2}} - 1 = (\nu + \frac{\sigma^2}{2})\partial t + O((\partial t)^2).$$

Here we used the identities

$$Ee^{\theta Z} = e^{\frac{\theta^2}{2}} \text{ and } e^x = 1 + x + O(x^2)$$

as $x \to 0$.

Similarly,

$$ER(0, \partial t)^2 = E(e^{G(0,\partial t)} - 1)^2 = Ee^{2G(0,\partial t)} - 2Ee^{G(0,\partial t)} + 1 =$$

$$e^{2\nu\partial t}Ee^{2\sigma\sqrt{\partial t}Z} - 2e^{\nu\partial t}Ee^{\sigma\sqrt{\partial t}Z} + 1 = e^{2\nu\partial t}e^{\frac{4\sigma^2\partial t}{2}} - 2e^{\nu\partial t}e^{\frac{\sigma^2\partial t}{2}} + 1 =$$

$$1 + (2\nu + \frac{4\sigma^2}{2})\partial t - 2(1 + (\nu + \frac{\sigma^2}{2})\partial t) + 1 + O((\partial t)^2) = \sigma^2\partial t + O((\partial t)^2).$$

The second identity now follows from the identity

$$\mathrm{Var}(R(0,\partial t)) = ER(0,\partial t)^2 - (ER(0,\partial t))^2$$

and the fact that $ER(0,\partial t) = O(\partial t)$.

**4.13.**

| AFGX | AZN | LME | HM | SDIA | SKA |
|---|---|---|---|---|---|
| $(0.06, 0.44)$ | $(-0.02, 0.49)$ | $(0.17, 0.97)$ | $(0.12, 0.79)$ | $(0.27, 1.06)$ | $(-0.02, 0.43)$ |

**4.14.** By Exercise 4.2b

$$E\frac{G(0,T)}{T} = \nu \text{ and } \mathrm{Var}(\frac{G(0,T)}{T}) = \frac{\sigma^2}{T}.$$

Therefore by Chebychev's inequality

$$P(|\frac{G(0,T)}{T} - \nu| > \epsilon) \le \frac{\sigma^2}{T\epsilon^2} \to 0$$

as $T \to \infty$ for each $\epsilon > 0$.

**4.15.** (a) Expected return $1/6 > 0$, drift $\frac{1}{2}\ln(2/3) < 0$. (b) Expected return $p/6$, $\max = 1/6$ is attained for $p = 1$. Drift $\frac{1}{2}\ln(1 + p/3 - 2p^2/3)$, $\max = \frac{1}{2}\ln(1 + 1/24) > 0$ is attained for $p = 1/4$.

**4.16.** We have

$$\partial S_t = S_t R(t, t + \partial t) = S_t(e^{\nu\partial t + \sigma\sqrt{\partial t}Z} - 1).$$

It follows from Exercise 4.12 that

$$E_t\partial S_t = S_t((\nu + \frac{\sigma^2}{2})\partial t + O((\partial t)^2)) \text{ and } E_t(\partial S_t)^2 = S_t^2(\sigma^2\partial t + O((\partial t)^2)).$$

We also have

$$E_t(\partial S_t)^k = S_t^k E(e^{\nu\partial t + \sigma\sqrt{\partial t}Z} - 1)^k = S_t^k E(\sigma\sqrt{\partial t}Z + O(\partial t))^k =$$

$$S_t^k E(\sigma^k \sqrt{\partial t}^k Z^k + O(\sqrt{\partial t}^{k-1}\partial t)) = S_t^k(\sigma^k(\partial t)^k EZ^k + O(\sqrt{\partial t}^{k+1})).$$

The expression for the expectations now follow since $EZ^k = 0$ for $k$ odd.

It follows that the standard deviation of $\partial S_t$ is of the order $\sqrt{\partial t}$, and hence $\partial S_t$ is of the same order.

Both the expected value and the standard deviation of $(\partial S_t)^2$ is of the order $\partial t$, and hence also $(\partial S_t)^2$ is of this order.

## Chapter 5

**5.1.**

$$\int_D^\infty e^{\sigma z}\phi(z)\,dz = \int_D^\infty e^{\sigma z - \frac{z^2}{2}} \frac{dz}{\sqrt{2\pi}} = e^{\frac{\sigma^2}{2}} \int_D^\infty e^{-\frac{(z-\sigma)^2}{2}} \frac{dz}{\sqrt{2\pi}} =$$

$$e^{\frac{\sigma^2}{2}} \int_{D-\sigma}^\infty e^{-\frac{z^2}{2}} \frac{dz}{\sqrt{2\pi}} = e^{\frac{\sigma^2}{2}}(1-\Phi(D-\sigma)).$$

Let $D = \frac{\ln A}{\sigma}$, then

$$E\max(0, e^{\sigma z} - A) = \int_D^\infty (e^{\sigma z} - A)\phi(z)\,dz =$$

$$\int_D^\infty e^{\sigma z}\phi(z)\,dz - A\int_D^\infty \phi(z)\,dz = e^{\frac{\sigma^2}{2}}(1-\Phi(D-\sigma)) - A(1-\Phi(D)).$$

Here we used the first identity.

**5.2.** By the put-call parity $P_0 = C_0 + \bar{K} - S_0$. Replace $C_0$ by the expression to the right in (5.7). This gives

$$P_0 = -S_0(1-\Phi(d_+)) + \bar{K}(1-\Phi(d_-)).$$

Identity (5.11) now follows from the identity $1-\Phi(d) = \Phi(-d)$ valid for all real $d$.

**5.3.**

$$s\phi(\frac{l}{\sigma} + \frac{\sigma}{2}) = se^{-\frac{1}{2}((\frac{l}{\sigma})^2 + l + (\frac{\sigma}{2})^2)}/\sqrt{2\pi} = se^{-\frac{l}{2}}C = \sqrt{sk}C,$$

where

$$C = e^{-\frac{1}{2}((\frac{l}{\sigma})^2 + (\frac{\sigma}{2})^2)}/\sqrt{2\pi}.$$

In the same way we obtain

$$k\phi(\frac{l}{\sigma} - \frac{\sigma}{2}) = ke^{\frac{l}{2}}C = \sqrt{sk}C.$$

**5.4.** (a)

$$\frac{\partial^2 C}{\partial s^2} = \frac{\partial \Delta}{\partial s} = \phi(d_+)\frac{\partial d_+}{\partial s} = \phi(d_+)\frac{1}{s\sigma\sqrt{T-t}}.$$

Here we used (5.18).

(b)

$$\frac{\partial C}{\partial t} = \frac{\partial}{\partial t}(s\Phi(d_+) - \bar{K}\Phi(d_-)) = s\phi(d_+)\frac{\partial d_+}{\partial t} - \bar{K}\phi(d_-)\frac{\partial d_-}{\partial t} - \frac{\partial \bar{K}}{\partial t}\Phi(d_) =$$

$$\bar{K}\phi(d_-)(\frac{\partial d_+}{\partial t} - \frac{\partial d_-}{\partial t}) - r\bar{K}\Phi(d_-) =$$

$$\bar{K}\phi(d_-)\frac{\partial \bar{\sigma}}{\partial t} - r\bar{K}\Phi(d_-) = -\bar{K}(\phi(d_-)\frac{\sigma}{2\sqrt{T-t}} + r\Phi(d_-)).$$

Here we used the identities $s\phi(d_+) = \bar{K}\phi(d_-)$, that is, (5.17), and $d_+ - d_- = \sigma\sqrt{T-t}$.

**5.5.** (a)

$$\frac{\partial C}{\partial \sigma} = \frac{\partial}{\partial \sigma}(s\Phi(d_+) - \bar{K}\Phi(d_-)) = s\phi(d_+)\frac{\partial d_+}{\partial \sigma} - \bar{K}\phi(d_-)\frac{\partial d_-}{\partial \sigma} =$$

$$\bar{K}\phi(d_-)(\frac{\partial d_+}{\partial \sigma} - \frac{\partial d_-}{\partial \sigma}) = \bar{K}\phi(d_-)\sqrt{T-t}.$$

Here we used the same identities as in Exercise 5.5b.

(b)

$$\frac{\partial C}{\partial r} = \frac{\partial}{\partial r}(s\Phi(d_+) - \bar{K}\Phi(d_-)) = s\phi(d_+)\frac{\partial d_+}{\partial r} - \bar{K}\phi(d_-)\frac{\partial d_-}{\partial r} - \frac{\partial \bar{K}}{\partial r}\Phi(d_-) =$$

$$\bar{K}\phi(d_-)(\frac{\partial d_+}{\partial r} - \frac{\partial d_-}{\partial r}) + (T-t)\bar{K}\Phi(d_-) = (T-t)\bar{K}\Phi(d_-).$$

Here we used (5.17) again and that $\frac{\partial(d_+ - d_-)}{\partial r} = 0$.

**5.6.**

| $T-t$ | $\Delta$ | $s\Gamma$ | $\Theta/s$ | $\rho/s$ | $\vartheta/s$ |
|---|---|---|---|---|---|
| 1 | 0.58 | 0.99 | −0.10 | 0.42 | 0.40 |
| 1/4 | 0.54 | 2.00 | −0.20 | 0.12 | 0.20 |
| 1/50 | 0.51 | 7.05 | −0.58 | 0.01 | 0.06 |

**5.7.** We shall differentiate the identity

$$P = C + \bar{K} - S,$$

and then use the expression for the corresponding derivative of $C$. Here $\bar{K} = e^{r(t-T)}K$.

(a)

$$\frac{\partial P}{\partial s} = \frac{\partial C}{\partial s} - 1 = \Phi(d_+) - 1 = -\Phi(-d_+).$$

(b)

$$\frac{\partial^2 P}{\partial s^2} = \frac{\partial^2 C}{\partial s^2}.$$

(c)

$$\frac{\partial P}{\partial t} = \frac{\partial C}{\partial t} + r\bar{K} = \bar{K}(-r\Phi(d_-) - \frac{\sigma\phi(d_-)}{2\sqrt{T-t}} + r) = \bar{K}(r\Phi(-d_-) - \frac{\sigma\phi(d_-)}{2\sqrt{T-t}}).$$

(d)

$$\frac{\partial P}{\partial \sigma} = \frac{\partial C}{\partial \sigma}.$$

(e)

$$\frac{\partial P}{\partial r} = \frac{\partial C}{\partial r} + (t-T)\bar{K} = \bar{K}\Phi(d_-)(T-t) - (T-t)\bar{K} = -\bar{K}\Phi(-d_-)(T-t).$$

**5.8.** (a) 2.294; (b) $dC = 0.86dS$, $A = 4.47$.

**5.9.** (a) 0.824; (b) 30%.

**5.10.** (a) $x$ equals the delta of the call; (b) $\frac{1}{2}\Gamma(\partial S)^2 + \Theta\partial t$, where $\Gamma$ and $\Theta$ are the gamma and theta of the call.

**5.11.**

| Day | 0 | 1 | 2 | 3 |
|---|---|---|---|---|
| Portfolio | 0.82 | 1.09 | 1.22 | 1.09 |
| $\Delta$ | 0.55 | 0.65 | 0.68 | |

**5.12.** It follows from Theorem 5.1 that the stock holding equals the $\Delta$ of the portfolio, and the remainder is cash. A call with the strike price $H$ thus has the stock holding $sh_C = S\Phi(d_+(H))$ and the amount of cash $ac_C = -\bar{H}\Phi(d_-(H))$. The corresponding quantities for a put with strike price $K$ equals $sh_P = -S\Phi(-d_+(K))$ and $ac_P = \bar{K}\Phi(-d_-(K))$.

Let $sh_i$ and $ac_i$ denote the stock holdings and the amount of cash for the three portfolios, $i = 1, 2, 3$. Then

$$sh_1 = \frac{S + sh_P}{1+p}, \quad ac_1 = \frac{ac_P}{1+p}$$

$$sh_2 = S + sh_P - sh_C, \quad ac_2 = ac_P - ac_C$$

$$sh_3 = \frac{S - sh_C}{1-c}, \quad ac_3 = \frac{-ac_C}{1-c}.$$

The result follows from this and the identity $1 - \Phi(x) = \Phi(-x)$.

**5.13.**

$$d_+(K) = \frac{\ln(S/\bar{K})}{\bar{\sigma}} + \frac{\bar{\sigma}}{2} = (\frac{\ln(S/\sqrt{\bar{K}\bar{H}})}{\bar{\sigma}} + \frac{\bar{\sigma}}{2}) + \frac{\ln(S/\bar{K}) - \ln(S/\sqrt{\bar{K}\bar{H}})}{\bar{\sigma}} =$$

$$d_+(\sqrt{KH}) + \frac{\ln(H/K)}{2\bar{\sigma}}.$$

Change $H$ and $K$ to get the second identity. The third then follows from $d_+(K)^2 - d_-(H)^2 = (d_+(K) + d_-(H))(d_+(K) - d_-(H))$.

**5.14.** $\Phi(b) - \Phi(-b) = 2\Phi(b) - 1$, where $b = \frac{\ln(H/K)}{2\sigma\sqrt{T-t}}$.

**5.15.** By Taylor's formula

$$\partial F = \frac{\partial F}{\partial t}\partial t + \frac{\partial F}{\partial s}\partial s + \frac{1}{2}(\frac{\partial^2 F}{\partial t^2}(\partial t)^2 + 2\frac{\partial^2 F}{\partial t \partial s}\partial t \partial s + \frac{\partial^2 F}{\partial s^2}(\partial s)^2) + O(|\partial t|^3 + |\partial s|^3)$$

It follows from this and Exercise 4.16 that

$$\partial F = \frac{\partial F}{\partial t}\partial t + \frac{\partial F}{\partial s}\partial s + \frac{1}{2}\frac{\partial^2 F}{\partial s^2}(\partial s)^2) + O(|\sqrt{\partial t}|^3).$$

**5.16.** Note that $S_{t_{k-1}}$ and $Z_k$ are stochastically independent. Let $E_{t_h}$ stand for the expectation given $S_{t_1}, \ldots, S_{t_h}$.

(a) Assume that $j < k$. Then

$$E\epsilon_k\epsilon_j = EE_{t_{k-1}}\epsilon_k\epsilon_j = E\epsilon_j E_{t_{k-1}}\epsilon_k = E\epsilon_j e(t_{k-1}, S_{t_{k-1}})E_{t_{k-1}}(Z_k^2 - 1)\partial t =$$

$$E\epsilon_j e(t_{k-1}, S_{t_{k-1}})0 = 0.$$

(b)

$$E\epsilon_k^2 = EE_{t_{k-1}}e(t_{k-1}, S_{t_{k-1}})^2(Z_k^2 - 1)^2(\partial t)^2 =$$

$$Ee(t_{k-1}, S_{t_{k-1}})^2 E(Z_k^2 - 1)^2(\partial t)^2.$$

(c)

$$E(Z_k^2 - 1)^2 = EZ_k^4 - 2EZ_k^2 + 1 = 3 - 2 + 1 = 2.$$

**5.17.**

$$\int_{t_{k-1}}^{t_k} G(t)\,dt - G(t_{k-1})\partial t = \int_{t_{k-1}}^{t_k} G(t) - G(t_{k-1})\,dt = \int_{t_{k-1}}^{t_k} dt \int_{t_{k-1}}^{t} G'(s)\,ds =$$

$$\int_{t_{k-1}}^{t_k} G'(s)\,ds \int_s^{t_k} dt = \int_{t_{k-1}}^{t_k} (t_k - s)G'(s)\,ds.$$

Therefore

$$|\int_0^T G(t)\,dt - \sum_{k=1}^n G(t_{k-1})\partial t| = |\sum_{k=1}^n (\int_{t_{k-1}}^{t_k} G(t)\,dt - G(t_{k-1})\partial t)| =$$

$$|\sum_{k=1}^n \int_{t_{k-1}}^{t_k} (t_k - s)G'(s)\,ds| \le \sum_{k=1}^n \int_{t_{k-1}}^{t_k} |t_k - s||G'(s)|\,ds \le$$

$$\sum_{k=1}^n \partial t \int_{t_{k-1}}^{t_k} |G'(s)|\,ds = \partial t \int_0^T |G'(s)|\,ds.$$

**5.18.** We shall use the same notation as in the proof of Theorem 5.1. Black–Scholes equation holds for the present values

$$\frac{\partial \bar{F}(t,\bar{s})}{\partial t} + \frac{\bar{s}^2\sigma^2}{2}\frac{\partial^2(t,\bar{s})F}{\partial \bar{s}^2} = 0.$$

We have

$$\frac{\partial \bar{F}(t,\bar{s})}{\partial t} = \frac{\partial e^{-rt}F(t,e^{rt}\bar{s})}{\partial t} = -re^{-rt}F + e^{-rt}\frac{\partial F}{\partial t} + e^{-rt}\frac{\partial F}{\partial s}\frac{\partial s}{\partial t} =$$

$$e^{-rt}(-rF + \frac{\partial F}{\partial t} + \frac{\partial F}{\partial t} + rs\frac{\partial F}{\partial s}),$$

and

$$\frac{\partial \bar{F}(t,\bar{s})}{\partial \bar{s}} = e^{-rt}\frac{\partial F}{\partial s}\frac{\partial s}{\partial \bar{s}} = e^{-rt}\frac{\partial F}{\partial s}e^{rt} = \frac{\partial F}{\partial s}.$$

Therefore

$$\frac{\partial^2 \bar{F}(t,\bar{s})}{\partial \bar{s}^2} = \frac{\partial^2 F}{\partial s^2}\frac{\partial s}{\partial \bar{s}} = \frac{\partial^2 F}{\partial s^2}e^{rt}.$$

If we replace the derivatives of the present values in Black–Scholes equation by the rightmost expressions in the above equations and multiply by $e^{rt}$, we get (5.20).

**5.19.** The proof is the same as in Exercise 5.18. Just more indices.

**5.20.** It should be clear that $g$ is continuous in $(0,\infty)^m$. Also

$$g(ks_1,...,ks_m) = ks_m f(\frac{ks_1}{ks_m},...,\frac{ks_{m-1}}{ks_m}) =$$

$$ks_m f(\frac{s_1}{s_m},...,\frac{s_{m-1}}{s_m}) = kg(s_1,...,s_m).$$

**5.21.** We shall use the fact that

$$\int e^{\theta_1 x_1 + \theta_1 x_1} \phi_{Q\tau}(x_1, x_2)\, dx_1 dx_2 = \Psi(\theta_1, \theta_2),$$

where

$$\Psi(\theta_1, \theta_2) = \exp(\frac{\tau}{2}(\theta_1^2 q_{11} + 2\theta_1\theta_2 q_{12} + \theta_2^2 q_{22})).$$

The generating function of the function to the left equals

$$\int \exp(\theta_1 x_1 + \theta_1 x_1 - \frac{q_{11}}{2}\tau + x_1)\phi_{Q\tau}(x_1, x_2)\, dx_1 dx_2 =$$

$$e^{-\frac{q_{11}}{2}\tau} \int \exp((\theta_1 + 1)x_1 + \theta_1 x_1)\phi_{Q\tau}(x_1, x_2)\, dx_1 dx_2 =$$

$$e^{-\frac{q_{11}}{2}\tau}\Psi(\theta_1 + 1, \theta_2) == e^{\tau L}\Psi(\theta_1, \theta_2),$$

where

$$L = \theta_1 q_{11} + \theta_2 q_{12}.$$

The function to the right has the generating function

$$\int e^{\theta_1 x_1 + \theta_1 x_1} \phi_{Q\tau}(x_1 - q_{11}\tau, x_2 - Q_{12}\tau)\, dx_1 dx_2 =$$

$$e^{\tau L} \int \phi_{Q\tau}(y_1, y_2)\, dy_1 dy_2 = e^{\tau L}\Psi(\theta_1, \theta_2).$$

Here we made the substitution $y_1 = x_1 - q_{11}\tau$, $y_2 = x_2 - q_{12}\tau$.

The functions on both sides have thus the same moment generating function, and hence they are equal.

**5.22.** $(\Phi(d_{12}), \Phi(d_{21}))$ for max, and $(\Phi(-d_{12}), \Phi(-d_{21}))$ for min.

**5.23.** (a) $c = 1/(2\Phi(\frac{\omega_{12}}{2}\sqrt{T})) = 0.83$, $C = 1/(2\Phi(-\frac{\omega_{12}}{2}\sqrt{T})) = 1.26$.

(b) $\max(S_1(T), S_2(T)) > 1/c = 1.21$, and $\min(S_1(T), S_2(T)) > 1/C = 0.80$, respectively.

(c) The second is preferable if $c/C < S_2(T)/S_1(T) < C/c$. In particular: $(c/C, C/c) = (0.66, 1.52)$.

## Chapter 6

**6.1.**

$$R_P(t, t+\partial t) = \frac{P(t+\partial t) - P(t)}{P(t)} = \sum \frac{V_j(t+\partial t) - V_j(t)}{P(t)} =$$

$$\sum \frac{V_j(t)}{P(t)} \frac{V_j(t+\partial t) - V_j(t)}{V_j(t)} = \sum v_j(t) R_j(t, t+\partial t).$$

The last identity follows from the fact that the number of each asset is the same at $t + \partial t$ as at $t$.

**6.2.**

$$\sigma_P^2 = v^2\sigma_1^2 + (1-v)^2\sigma_2^2 + 2v(1-v)\sigma_1\sigma_2\rho$$

The derivative of $\sigma_P^2$ with respect to $v$ equals

$$2(v\sigma_1^2 - (1-v)\sigma_2^2 + (1-2v)\sigma_1\sigma_2\rho),$$

and this equals 0 for

$$v = \frac{\sigma_2^2 - \sigma_1\sigma_2\rho}{\sigma_1^2 + \sigma_2^2 - 2\sigma_1\sigma_2\rho} = 0.142$$

The minimum is attained for this value since $\sigma_P^2$ is a positive definite quadratic form, and a calculation yields that this minimum equals $0.068 = 0.261^2$.

**6.3.** We have

$$\sigma_P^2 = \sum_{i,j} \sigma_{i,j} v_i v_j,$$

where $v_i = C/\sigma_i$, $\sigma_{i,i} = \sigma_i^2$ and $\sigma_{i,j} = \sigma_i\sigma_j\rho$ for $i \neq j$. The constant $C$ is determined by the condition $v_1 + \cdots + v_m = 1$; $C = H/m$, where

$$H = \frac{m}{\frac{1}{\sigma_1} + \cdots + \frac{1}{\sigma_m}}$$

is the harmonic mean of the volatilities.

The following calculation gives the result

$$\sigma_P^2 = \sum_{i=j} + \sum_{i \neq j} = \sum_i \sigma_i^2 (\frac{C}{\sigma_i})^2 + \sum_{i \neq j} \sigma_i\sigma_j\rho \frac{C}{\sigma_i}\frac{C}{\sigma_j} =$$

$$C^2(\sum_i 1 + \sum_{i \neq j} \rho) = \frac{H^2}{m}(m + m(m-1)\rho) = H^2(\rho + \frac{1-\rho}{m}).$$

**6.4.** (a) $v_i = \frac{1}{\sigma_i^2}/S$, $\sigma_*^2 = 1/S$, where $S = \frac{1}{\sigma_1^2} + \frac{1}{\sigma_2^2} + \frac{1}{\sigma_3^2}$. (b) $(1/3, 1/3, 1/3)$, $\sigma_*^2 = 2/15$. (c) $(3/7, 2/7, 2/7), \sigma_*^2 = 3/35$.

**6.5.** Alternative b. (The volatilities are $13.5\%$ and $11.8\%$, respectively.)

**6.6.** $w = p\kappa$, where $\kappa = \frac{\sigma}{\tau}|\rho|$. The volatility equals $p\sigma\sqrt{1-\rho^2}$.

**6.7.** (a) $(0.50, 0.42, 0.08), 11.7\%$; (b) $(0.50, 2.53, -2.03), 8.9\%$; (c) $12.1\%$.

**6.8.** Put $\boldsymbol{u} = \boldsymbol{\mu} - \mu_{min}\mathbf{1}$. Then $\boldsymbol{v}_{aux} = Q^{-1}\boldsymbol{u}$, and

$$\sigma_{aux}^2 = Q^{-1}\boldsymbol{u} \cdot QQ^{-1}\boldsymbol{u} = Q^{-1}\boldsymbol{u} \cdot \boldsymbol{u} = \boldsymbol{u} \cdot Q^{-1}\boldsymbol{u}.$$

Also

$$\mu_{aux} = \mu \cdot Q^{-1}\boldsymbol{u} = \boldsymbol{u} \cdot Q^{-1}\boldsymbol{u} + \mu_{min}\mathbf{1} \cdot Q^{-1}\boldsymbol{u} = \boldsymbol{u} \cdot Q^{-1}\boldsymbol{u}$$

because

$$\mathbf{1} \cdot Q^{-1}\boldsymbol{u} = \mathbf{1} \cdot Q^{-1}\boldsymbol{\mu} - \mu_{min}\mathbf{1} \cdot Q^{-1}\mathbf{1} = 0.$$

**6.9.**

$$\mu_P = (1-a)r_f + a\mu_{min} + b\mu_{aux} = r_f + a(\mu_{min} - r_f) + b\sigma_{aux}^2.$$

Here we used (6.20).

$$\sigma_P^2 = (a\boldsymbol{v}_{min} + b\boldsymbol{v}_{aux}) \cdot Q(a\boldsymbol{v}_{min} + b\boldsymbol{v}_{aux}) =$$

$$a^2\boldsymbol{v}_{min} \cdot Q\boldsymbol{v}_{min} + b^2\boldsymbol{v}_{aux} \cdot Q\boldsymbol{v}_{aux} + 2ab\boldsymbol{v}_{min} \cdot Q\boldsymbol{v}_{aux} = a^2\sigma_{min}^2 + b^2\sigma_{aux}^2$$

because

$$\boldsymbol{v}_{min} \cdot Q\boldsymbol{v}_{aux} = \sigma_{min}^2\mathbf{1} \cdot Q^{-1}\boldsymbol{u} = 0.$$

Here $\boldsymbol{u}$ is as in the solution of Exercise 6.8.

The expression for the drift follows from the identity (6.16) and the identities (6.22) and (6.23).

**6.10.** Replace $\boldsymbol{u}$ in the derivation of $\sigma_{aux}$ in Exercise 6.8 by $\boldsymbol{\mu} - r_f\mathbf{1}$ to obtain (6.27). The expected return equals

$$\mu_{max} = (1-\omega)r_f + \boldsymbol{\mu} \cdot \boldsymbol{v}_{max} = (1-\omega)r_f + \boldsymbol{\mu} \cdot Q^{-1}(\boldsymbol{\mu} - r_f\mathbf{1}) =$$

$$(1-\omega)r_f + (\boldsymbol{\mu} - r_f\mathbf{1}) \cdot Q^{-1}(\boldsymbol{\mu} - r_f\mathbf{1}) + r_f\mathbf{1} \cdot Q^{-1}(\boldsymbol{\mu} - r_f\mathbf{1}) = (1-\omega)r_f + \sigma_{max}^2 + r_f\omega.$$

The identity (6.28) now follows from (6.16).

**6.11.**

$$Q^{-1}(\boldsymbol{\mu} - r_f\mathbf{1}) = Q^{-1}(\boldsymbol{\mu} - \mu_{min}\mathbf{1}) + Q^{-1}(\mu_{min}\mathbf{1} - r_f\mathbf{1}) = \boldsymbol{v}_{aux} + \omega\boldsymbol{v}_{min}.$$

**6.12.**

$$\sigma_{\mathrm{Mz}}^2 = \frac{\boldsymbol{v}_{max} \cdot Q\boldsymbol{v}_{max}}{\omega^2} = \frac{\sigma_{max}^2}{\omega^2}.$$

$$\mu_{\mathrm{Mz}} = \frac{\boldsymbol{\mu} \cdot \boldsymbol{v}_{max}}{\omega} = \frac{\sigma_{max}^2 + r_f\omega}{\omega}.$$

The last identity follows from the calculation in the solution of Exercise 6.10.

**6.13.** Put $\kappa = \sigma/\sigma_{max}$. It follows from (6.29) that the weights are of the form (6.21) with $a = \omega\kappa$ and $b = \kappa$. It therefore follows from (6.22) that the expected return of the portfolio equals

$$\mu_P = r_f + \omega^2\kappa\sigma_{min}^2 + \kappa\sigma_{aux}^2,$$

and from (6.22) that

$$\sigma_P^2 = (\omega\kappa)^2\sigma_{min}^2 + \kappa^2\sigma_{aux}^2.$$

Using (6.30), we now get

$$\mu_P = r_f + \kappa\sigma_{max}^2 = r_f + \sigma\sigma_{max}, \text{ and } \sigma_P^2 = \kappa^2\sigma_{max}^2 = \sigma^2.$$

The first identity now follows from (6.16), and the second one follows from (6.28).

**6.14.** The matrix $Q$ is nonsingular. Therefore $\boldsymbol{v} = \boldsymbol{v}'$ if and only if $Q\boldsymbol{v} = Q\boldsymbol{v}'$, that is,

$$\frac{\boldsymbol{\mu}' - r_f\mathbf{1}}{\sigma'_{max}} = \frac{\boldsymbol{\mu} - r_f\mathbf{1}}{\sigma_{max}}.$$

Assume $\boldsymbol{v} = \boldsymbol{v}'$, then $\boldsymbol{\mu}' - r_f\mathbf{1} = a(\boldsymbol{\mu} - r_f\mathbf{1})$ with $a = \sigma'_{max}/\sigma_{max}$.

Conversely, if this identity holds for some $a$, then

$$Q\boldsymbol{v}' = \frac{a\sigma_{max}}{\sigma'_{max}}Q\boldsymbol{v}.$$

The identity (6.27) implies

$$\sigma'_{max} = a\sigma_{max}$$

and hence $Q\boldsymbol{v}' = Q\boldsymbol{v}$.

**6.15.**

$$\boldsymbol{a} \cdot \boldsymbol{b} = \mathbf{1} \cdot Q^{-1}(\boldsymbol{\mu} - r_f\mathbf{1}) = \mathbf{1} \cdot Q^{-1}\boldsymbol{\mu} - r_f\mathbf{1} \cdot Q^{-1}\mathbf{1} = \omega.$$

The last identity follows from (6.18) and (6.8). Also

$$|\boldsymbol{d}|^2 = |\boldsymbol{a}|^2 - \frac{(\boldsymbol{a} \cdot \boldsymbol{b})^2}{|\boldsymbol{b}|^2}.$$

It follows from (6.27) that $|a|^2 = \sigma_{max}^2$ and from (6.8) that $1/|b|^2 = \sigma_{min}^2$. The first identity of the exercise now follows from (6.30).

**6.16.** We have $\boldsymbol{v} = \boldsymbol{v}_{min} + k\boldsymbol{v}_{aux}$ and therefore

$$r = \mu_{min} + k\mu_{aux} = \mu_{min} + k\sigma^2_{aux}.$$

**6.17.** Write $(\sigma_T, r_T)$ for the tangent point.

The tangent line has the form $\sigma = l(r - r_f)$, and hence the slope $l$ satisfies

$$(1)\ \ \sigma_T = l(r_T - r_f).$$

Differentiating both sides of (6.43) with respect to $r$, we get

$$2\sigma\frac{\partial\sigma}{r} = 2(r - \mu_{min})/\sigma^2_{aux},$$

and hence

$$(2)\ \ l = \frac{r_T - \mu_{min}}{\sigma_T\sigma^2_{aux}}.$$

The third equation is (6.43) with $(\sigma, r) = (\sigma_T, r_T)$.

We thus have three equations and three unknown quantities: $l$, $\sigma_T$, and $r_T$.

(1) gives

$$(4)\ l = \frac{\sigma_T}{r - r_f}.$$

This and (2) then gives

$$(5)\ \ \sigma_T^2 = \frac{(r_T - r_f)(r_T - \mu_{min})}{\sigma^2_{aux}}.$$

The latter gives together with the third equation

$$\frac{(r_T - r_f)(r_T - \mu_{min})}{\sigma^2_{aux}} = \sigma^2_{min} + \frac{(r_T - \mu_{min})^2}{\sigma^2_{aux}}.$$

That is,

$$r_T = \mu_{min} + \frac{\sigma^2_{min}\sigma^2_{aux}}{\mu_{min} - r_f} = \mu_{min} + \frac{\sigma^2_{aux}}{\omega}.$$

This gives the second statement of the exercise.

We also have

$$r_T - r_f = \omega\sigma^2_{min} + \frac{\sigma^2_{aux}}{\omega} = \frac{\sigma^2_{max}}{\omega}.$$

Here we used (6.30). This together with (5) yields

$$\sigma_T^2 = \frac{\frac{\sigma_{max}^2}{\omega}\frac{\sigma_{aux}^2}{\omega}}{\sigma_{aux}^2} = \frac{\sigma_{max}^2}{\omega^2}.$$

The equation (4) now yields

$$l = \frac{\sigma_{max}}{\omega}\frac{\omega}{\sigma_{max}^2} = \frac{1}{\sigma_{max}}.$$

Therefore the line (6.42) and the tangent line are the same.

**6.18.** (a) 4000, 208, 2133. (b) 0.14, 0.42, 0.44. Buy 1816, −35, and 37 stocks, respectively.

**6.19.**

$$P(t_n) = P(t_0)\prod_{k=1}^{n}\frac{P(t_k)}{P(t_{k-1})} = P(t_0)\prod_{k=1}^{n}(1 + R_P(t_{k-1}, t_k)).$$

See Exercise 6.1 for the remaining details.

## Chapter 7

**7.1.**

$$\text{Var}(R - bR_M) = \text{Var}(R) - 2b\text{Cov}(R, R_M) + b^2\text{Var}(R_M) =$$
$$\text{Var}(R) - \beta^2\text{Var}(R_M) + \text{Var}(R_M)(b-\beta)^2.$$

**7.2.** We have

$$\text{Cov}(R_M, e) = \text{Cov}(R, R_M) - \beta\text{Var}(R_M) = 0,$$

and hence $R_M$ and $e$ are uncorrelated. Therefore

$$\text{Var}(R) = \text{Var}(\beta R_M + e) = \text{Var}(\beta R_M) + \text{Var}(e).$$

The first of the remaining identities is a fairly direct consequence of of the definitions of $\beta$ and $\rho$. The last identity then follows from this and the first identity.

**7.3.**

$$\sum p_i\beta_i = \sum p_i\text{Cov}(R_i, R_M)/\sigma_M^2 = \text{Cov}(\sum p_i R_i, R_M)/\sigma_M^2 =$$
$$\text{Cov}(R_M, R_M)/\sigma_M^2 = 1.$$

**7.4.** Let $\boldsymbol{R} = (R_1, ..., R_m)$ denote the returns of the assets, and $\sigma_P$ the volatility of the portfolio. Then $R_P = \boldsymbol{v} \cdot \boldsymbol{R}$, and

$$\sigma_P^2 = \text{Var}(R_P) = \boldsymbol{v} \cdot Q\boldsymbol{v}.$$

Also

$$\rho(\boldsymbol{v}) = \frac{\text{Cov}(R_P, R_M)}{\sigma_P \sigma_M} = \frac{\sum v_i \text{Cov}(R_i, R_M)}{\sigma_P \sigma_M} = \frac{\sum v_i \beta_i \sigma_M^2}{\sigma_P \sigma_M} = \frac{\boldsymbol{v} \cdot \boldsymbol{\beta}}{\sigma_P} \sigma_M.$$

**7.5.** (a) The condition $\mathbf{1} \cdot \boldsymbol{v} = 1$ gives

$$c = 1/\mathbf{1} \cdot Q^{-1}\boldsymbol{\beta},$$

and hence (7.2) holds.

The volatility satisfies

$$\sigma_\beta^2 = \boldsymbol{v}_\beta \cdot Q\boldsymbol{v}_\beta = c^2(Q^{-1}\boldsymbol{\beta}) \cdot Q(Q^{-1}\boldsymbol{\beta}) = c^2\boldsymbol{\beta} \cdot (Q^{-1}QQ^{-1})\boldsymbol{\beta} = c^2\boldsymbol{\beta} \cdot Q^{-1}\boldsymbol{\beta}.$$

The expression for the volatility follows from this.

The portfolio has the return

$$R_\beta = \boldsymbol{v}_\beta \cdot \boldsymbol{R} = \sum v_\beta(i) R_i.$$

Therefore

$$\beta_\beta = \text{Cov}(\sum v_\beta(i) R_i, R_M)/\sigma_M^2 = \sum v_\beta(i)\text{Cov}(R_i, R_M)/\sigma_M^2 =$$

$$\sum v_\beta(i)\beta_i\sigma_M^2/\sigma_M^2 = \boldsymbol{v}_\beta \cdot \boldsymbol{\beta} = c\boldsymbol{\beta} \cdot Q^{-1}\boldsymbol{\beta}.$$

(b)

$$\rho_{\beta,M} = \frac{\text{Cov}(R_P, R_M)}{\sigma_\beta \sigma_M} = \frac{\beta_\beta \sigma_M^2}{\sigma_\beta \sigma_M}.$$

The expression for the correlation now follows from this and the expressions for $\sigma_\beta$ and $\beta_\beta$ given in 7.5a.

**7.6.** (a)

$$\rho_{min,M} = \frac{\text{Cov}(\boldsymbol{v}_{min} \cdot \boldsymbol{R}, R_M)}{\sigma_{min}\sigma_M} = \frac{\sum v_i \text{Cov}(R_i, R_M)}{\sigma_P \sigma_M} =$$

$$\frac{\sum v_i \beta_i \sigma_M^2}{\sigma_P \sigma_M} = \frac{\boldsymbol{v} \cdot \boldsymbol{\beta}}{\sigma_P} \sigma_M.$$

(b) The portfolio has the return

$$R_\beta = \boldsymbol{v}_{min} \cdot \boldsymbol{R} = \sum v_{min}(i) R_i.$$

We get in the same way as in Exercise 7.5 that

$$\beta_{min} = \boldsymbol{v}_{min} \cdot \boldsymbol{\beta} = \sigma^2_{\min} \mathbf{1} \cdot Q^{-1}\boldsymbol{\beta}.$$

**7.7.** By (7.4) and (7.3)

$$\beta_\beta = \sigma^2_\beta \mathbf{1} \cdot Q^{-1}\boldsymbol{\beta}.$$

The result now follows from (7.5) and the fact that $\sigma_\beta = \sigma_{min}$ only if the beta portfolio has the same weights as the minimum-variance portfolio.

An alternative solution can be obtained by applying Schwarz inequality, $(\boldsymbol{x} \cdot \boldsymbol{y})^2 \leq |\boldsymbol{x}|^2 |\boldsymbol{y}|^2$, with $\boldsymbol{x} = Q^{-1/2}\mathbf{1}$ and $\boldsymbol{y} = Q^{-1/2}\boldsymbol{\beta}$.

**7.8.**

$$\text{Cov}(R_P, R_\beta) = \sum_{i=1}^m v_i \text{Cov}(R_i, R_\beta) = \sum_{i=1}^m v_i \sum_{j=1}^m v_\beta(j) \text{Cov}(R_i, R_j) =$$

$$\boldsymbol{v} \cdot Q\boldsymbol{v}_\beta = \boldsymbol{v} \cdot \boldsymbol{\beta}/\mathbf{1} \cdot Q^{-1}\boldsymbol{\beta} = \text{Cov}(R_P, R_M)/(\sigma^2_M \mathbf{1} \cdot Q^{-1}\boldsymbol{\beta}).$$

**7.9.** We shall minimize $\boldsymbol{v} \cdot Q\boldsymbol{v}$ (or equivalently $\boldsymbol{v} \cdot Q\boldsymbol{v}/2$) subject to the constraints $\boldsymbol{v} \cdot \mathbf{1} = 1$ and $\boldsymbol{v} \cdot \boldsymbol{\beta} = 0$.

Using Lagrange multipliers, we get the equations

$$Q\boldsymbol{v} = \lambda\mathbf{1} + \boldsymbol{\mu}\boldsymbol{\beta}, \quad \boldsymbol{v} \cdot \mathbf{1} = 1, \quad \boldsymbol{v} \cdot \boldsymbol{\beta} = 0.$$

Therefore

$$\boldsymbol{v} = \lambda Q^{-1}\mathbf{1} + \mu Q^{-1}\boldsymbol{\beta} = \lambda' \boldsymbol{v}_{min} + \mu' \boldsymbol{v}_\beta.$$

Here we used (6.9) and (7.2). The constraints give

$$\lambda' + \mu' = 1, \text{ and } \lambda' \beta_{min} + \mu' \beta_\beta = 0.$$

Therefore $\mu' = 1 - \lambda'$ and

$$\lambda' = \frac{\beta_\beta}{\beta_\beta - \beta_{min}} = \frac{\sigma^2_\beta}{\sigma^2_\beta - \sigma^2_{min}} = 1 + \gamma.$$

Here we used (7.5), (7.4), and (7.3).

**7.10.** We have $\boldsymbol{\mu} = r_f \mathbf{1} + \Delta\boldsymbol{\beta}$, and hence

$$\mu_P = \boldsymbol{v} \cdot (r_f \mathbf{1} + \Delta\boldsymbol{\beta}) = r_f \boldsymbol{v} \cdot \mathbf{1} + \Delta \boldsymbol{v} \cdot \boldsymbol{\beta} = r_f + \Delta \boldsymbol{v} \cdot \boldsymbol{\beta}.$$

**7.11.** We have $\boldsymbol{\mu} - r_f \mathbf{1} = \Delta \boldsymbol{\beta}$, and hence

$$\boldsymbol{v}_{Mz} = Const.\, Q^{-1}(\boldsymbol{\mu} - r_f \mathbf{1}) = Const.\, Q^{-1}(\Delta \boldsymbol{\beta}) = Const.\, Q^{-1}\boldsymbol{\beta} = \boldsymbol{v}_{\beta}.$$

The last identity follows from the fact that both $\boldsymbol{v}_{Mz}$ and $\boldsymbol{v}_{\beta}$ sum to 1.

**7.12.** (a) We have $\boldsymbol{\mu} - r_f \mathbf{1} = \Delta \boldsymbol{\beta}$, and hence by (6.26)

$$\boldsymbol{v}_{max} = Q^{-1}(\boldsymbol{\mu} - r_f \mathbf{1}) = Q^{-1}(\Delta \boldsymbol{\beta}) = \Delta Q^{-1}\boldsymbol{\beta},$$

by (6.27)

$$\sigma^2_{max} = (\boldsymbol{\mu} - r_f \mathbf{1}) \cdot Q^{-1}(\boldsymbol{\mu} - r_f \mathbf{1}) = (\Delta \boldsymbol{\beta}) \cdot Q^{-1}(\Delta \boldsymbol{\beta}) = \Delta^2 \boldsymbol{\beta} \cdot Q^{-1}\boldsymbol{\beta},$$

and by (7.17), (6.25), and (7.5)

$$\omega = \frac{\mu_{min} - r_f}{\sigma^2_{min}} = \frac{\Delta \beta_{min}}{\sigma^2_{min}} = \Delta \mathbf{1} \cdot Q^{-1}\boldsymbol{\beta}.$$

(b) This follows from 7.12a.

(c) The first identity follows from (6.29) and the identities in Exercise 7.12a. The second identity follows from (6.20) and the fact that $\boldsymbol{\mu} - \mu_{min}\mathbf{1} = \Delta(\boldsymbol{\beta} - \beta_{min}\mathbf{1})$.

**7.13.**

$$\beta_i = \frac{\mathrm{Cov}(R_i, R_M)}{\sigma^2_M} = \frac{\sum_j p_j \mathrm{Cov}(R_i, R_j)}{\sigma^2_M} = \frac{(Q\boldsymbol{p})_i}{\sigma^2_M}.$$

Therefore

$$Q^{-1}\boldsymbol{\beta} = \frac{\boldsymbol{p}}{\sigma^2_M}, \tag{A.1}$$

and hence

$$\mathbf{1} \cdot Q^{-1}\boldsymbol{\beta} = \frac{1}{\sigma^2_M}. \tag{A.2}$$

**7.14.** Identity (A.1) implies (A.2), and

$$\boldsymbol{\beta} \cdot Q^{-1}\boldsymbol{\beta} = \frac{\boldsymbol{\beta} \cdot \boldsymbol{p}}{\sigma^2_M} = \frac{1}{\sigma^2_M}.$$

Here we used Exercise 7.3. The result now follows from (7.5) and (7.18).

## Chapter 8

**8.1.**

$$E[G_A(0,1) - G_B(0,1)] = EG_A(0,1) - EG_B(0,1) = \nu_A - \nu_B.$$

$$\text{Var}(G_A(0,1) - G_B(0,1)) =$$

$$\text{Var}(G_A(0,1)) - 2\text{Cov}(G_A(0,1), G_B(0,1)) + \text{Var}(G_B(0,1)) =$$

$$\sigma_A^2 - 2\sigma_A\sigma_B\rho + \sigma_B^2.$$

**8.2.** Let

$$\sigma^2 = \text{Var}(G_A(0,T) - G_B(0,T))/T,$$

and put

$$U = \frac{G_A(0,T) - G_B(0,T)}{T\hat{\sigma}}, \quad V = \frac{\hat{\sigma}}{\sigma}.$$

Then $\hat{P}_r = U/V$, $U$ and $V$ are independent, $U$ is $N(P_r, 1/T)$, and $V$ is asymptotically $N(1, 1/(2n))$. Therefore

$$\hat{P}_r = \frac{P_r + \frac{Z_1}{\sqrt{T}}}{1 + \frac{Z_2}{\sqrt{2n}}} = \frac{P_r + \frac{Z_1}{\sqrt{T}}}{1 + Z_2\sqrt{\frac{\partial t}{2T}}},$$

where $Z_1$ and $Z_2$ are independent and asymptotically $N(0,1)$. It follows that

$$\hat{P}_r = P_r + \frac{Z_1}{\sqrt{T}} - Z_2\sqrt{\frac{\partial t}{2T}} + O(\frac{\sqrt{\partial t}}{T})$$

from which the result follows.

**8.3.** Let $\boldsymbol{u} = (\boldsymbol{v}_A - \boldsymbol{v}_B)/2$, $\boldsymbol{w} = (\boldsymbol{v}_A + \boldsymbol{v}_B)/2$. Then $\boldsymbol{v}_A = \boldsymbol{u} + \boldsymbol{w}$, $\boldsymbol{v}_B = -\boldsymbol{u} + \boldsymbol{w}$. Using (8.6), we get

$$G_A(0,1) - G_B(0,1) =$$

$$(\gamma_A - \gamma_B)r_f + (\boldsymbol{v}_A - \boldsymbol{v}_B) \cdot \boldsymbol{\mu} - \frac{1}{2}(\boldsymbol{v}_A \cdot Q\boldsymbol{v}_A - \boldsymbol{v}_B \cdot Q\boldsymbol{v}_B) + (\boldsymbol{v}_A - \boldsymbol{v}_B) \cdot \boldsymbol{X}(1) =$$

$$(\gamma_A - \gamma_B)r_f + 2\boldsymbol{u} \cdot \boldsymbol{\mu} - \frac{1}{2}((\boldsymbol{u}+\boldsymbol{w}) \cdot Q(\boldsymbol{u}+\boldsymbol{w}) - (-\boldsymbol{u}+\boldsymbol{w}) \cdot Q(-\boldsymbol{u}+\boldsymbol{w})) + 2\boldsymbol{u} \cdot \boldsymbol{X}(t) =$$

$$(\gamma_A - \gamma_B)r_f + 2\boldsymbol{u} \cdot \boldsymbol{\mu} - \frac{1}{2}4\boldsymbol{u} \cdot Q\boldsymbol{w} + 2\boldsymbol{u} \cdot \boldsymbol{X}(1) =$$

$$(\gamma_A - \gamma_B)r_f + 2\boldsymbol{u} \cdot (\boldsymbol{\mu} - Q\boldsymbol{w}) + 2\boldsymbol{u} \cdot \boldsymbol{X}(1).$$

The random variable $\boldsymbol{X}(1)$ has zero expectation. The drift therefore equals

$$(\gamma_A - \gamma_B)r_f + 2\boldsymbol{u} \cdot (\boldsymbol{\mu} - Q\boldsymbol{w}).$$

Also

$$\mathrm{Var}(G_A(0,1) - G_B(0,1)) = \mathrm{Var}(2\boldsymbol{u} \cdot \boldsymbol{X}(1)) = (2\boldsymbol{u}) \cdot Q(2\boldsymbol{u}).$$

The result follows from these identities.

**8.4.** We shall apply (8.7) with $\gamma_A = 1-a, \gamma_B = 1-b, \boldsymbol{v}_A = a\boldsymbol{v}_{Mz}$, and $\boldsymbol{v}_B = b\boldsymbol{v}_{Mz}$:

$$P_r(A,B) = \frac{(\gamma_A - \gamma_B)r_f + (\boldsymbol{v}_A - \boldsymbol{v}_B) \cdot (\boldsymbol{\mu} - \frac{1}{2}Q(\boldsymbol{v}_A + \boldsymbol{v}_B))}{\sqrt{(\boldsymbol{v}_A - \boldsymbol{v}_B) \cdot Q(\boldsymbol{v}_A - \boldsymbol{v}_B)}} =$$

$$\frac{(b-a)r_f + (a-b)\boldsymbol{v}_{Mz} \cdot (\boldsymbol{\mu} - \frac{a+b}{2}Q\boldsymbol{v}_{Mz})}{|a-b|\sqrt{\boldsymbol{v}_{Mz} \cdot Q\boldsymbol{v}_{Mz}}} =$$

$$\frac{(b-a)r_f + (a-b)(r_f + \frac{\sigma_{max}^2}{\omega} - \frac{a+b}{2}\frac{\sigma_{max}^2}{\omega^2})}{|a-b|\frac{\sigma_{max}}{\omega}} =$$

$$\frac{(a-b)(\frac{\sigma_{max}^2}{\omega} - \frac{a+b}{2}\frac{\sigma_{max}^2}{\omega^2})}{|a-b|\frac{\sigma_{max}}{\omega}} =$$

$$\mathrm{sign}(a-b)(\sigma_{max} - \frac{a+b}{2}\frac{\sigma_{max}}{\omega}).$$

Here we used (6.32) to obtain the third identity.

**8.5.**

$$Sh = \frac{\mu - r_f}{\sigma} = \frac{\nu + \frac{\sigma^2}{2} - r_f}{\sigma} = Sh + \frac{\sigma}{2}.$$

**8.6.** Performance ratios: $0.76, 1.33, 1.25, 0.78$.

Sharpe ratios if $\mu = e^\nu - 1$: $0.87, 1.58, 1.63, 1.33$.

**8.7.**

$$Sh_i = \frac{\mu_i - r_f}{\sigma_i} = \frac{\beta_i(\mu_M - r_f)}{\sigma_i} = \rho_i \frac{\mu_M - r_f}{\sigma_M} = \rho_i Sh_M.$$

**8.8.** $2.3\%, 1.9\%, 3.5\%, 11.8\%$.

**8.9.** In this case the $m = 1$, and $S(0) = P(0)$. Therefore

$$\sigma_P = |w(0)|\sigma = |\Delta_P|\sigma.$$

It follows from Exercise 5.7a that the delta for the put equals $-\Phi(-d_+(K))$, and from (5.18) that the delta for the call equals $\Phi(d_+(H))$. Therefore $\Delta_P$ for the three portfolios equal

$$\frac{1 - \Phi(-d_+(K))}{1+p}, \quad 1 - \Phi(-d_+(K)) - \Phi(-d_+(H)), \quad \text{and} \quad \frac{1 - \Phi(d_+(H))}{1-c}.$$

The result now follows from the identity $1 - \Phi(-d) = \Phi(d)$.

**8.10.** The function $F$ is decreasing because

$$\frac{d}{d\delta}F(\delta) = \frac{d}{d\delta}E\min(0,\delta+Y)^2 = E\frac{d}{d\delta}\min(0,\delta+Y)^2 = E2\min(0,\delta+Y) \leq 0.$$

Let $\delta > 0$. Then

$$F(\delta) = \int \min(0,\delta+y)^2 dF(y) = \int_{y<-\delta} (|y|-\delta)^2 dF(y) \leq \int_{y<-\delta} y^2 dF(y) \to 0,$$

as $\delta \to \infty$.

**8.11.** Using the identity $Y^2 = \min(0,Y)^2 + \min(0,-Y)^2$, we get

$$1 = EY^2 = E\min(0,Y)^2 + E\min(0,-Y)^2 = 2E\min(0,Y)^2 = 2F(0).$$

**8.12.**

$$F(0) = \int_0^\infty \min(0,x-1)^2 e^{-x} dx = \int_0^1 (x-1)^2 e^{-x} dx = \int_0^1 y^2 e^y dx/e =$$

$$[(y^2-2y+2)e^y]_0^1/e = (e-2)/e.$$

Here we made the substitution $y = x - 1$ and two partial integrations.

**8.13.**

$$F(\delta) = \int_{-\infty}^\infty \min(0,\delta+y)^2 \phi(y)\,dy = \int_{-\infty}^{-\delta} (\delta+y)^2 \phi(y)\,dy =$$

$$\int_{-\infty}^{-\delta} \delta^2 \phi(y)\,dy + \int_{-\infty}^{-\delta} 2\delta y \phi(y)\,dy + \int_{-\infty}^{-\delta} y^2 \phi(y)\,dy.$$

Clearly,

$$\int_{-\infty}^{-\delta} \delta^2 \phi(y)\,dy = \delta^2 \Phi(-\delta).$$

Using the identity $\phi'(y) = -y\phi(y)$, we get

$$\int_{-\infty}^{-\delta} 2\delta y \phi(y)\,dy = 2\delta[-\phi(y)]_{-\infty}^{-\delta} = -2\delta\phi(-\delta).$$

By a partial integration, we obtain

$$\int_{-\infty}^{-\delta} y^2 \phi(y)\,dy = [-y\phi(y)]_{-\infty}^{-\delta} + \int_{-\infty}^{-\delta} \phi(y)\,dy = \delta\phi(-\delta) + \Phi(-\delta).$$

The result now follows from the fact that $\phi(-\delta) = \phi(\delta)$.

**8.14.** We shall use the fact that the identity

$$J\boldsymbol{z} = (\boldsymbol{z} \cdot \mathbf{1})\mathbf{1}$$

holds for all $\boldsymbol{z}$.

Assume $\boldsymbol{x} = (I - \frac{1}{m}J)\boldsymbol{y}$. Then

$$\mathbf{1} \cdot \boldsymbol{x} = \mathbf{1} \cdot (I - \frac{1}{m}J)\boldsymbol{y} = (I - \frac{1}{m}J)\mathbf{1} \cdot \boldsymbol{y} = (\mathbf{1} - \frac{1}{m}J\mathbf{1}) \cdot \boldsymbol{y} = 0.$$

Conversely, if $\mathbf{1} \cdot \boldsymbol{x} = 0$. Then $J\boldsymbol{x} = 0$, and hence

$$\boldsymbol{x} = (I - \frac{1}{m}J)\boldsymbol{x}.$$

**8.15.** In the first case we have $\nabla F(\boldsymbol{v}) = Q\boldsymbol{v}$, and in the second case we have

$$\nabla F(\boldsymbol{v}) = \frac{1}{n\partial t}\sum_{j=1}^{n} \min(0, \boldsymbol{v} \cdot \boldsymbol{R}_j - r_f \partial t)\boldsymbol{R}_j.$$

The identities therefore follows from the identity $J\boldsymbol{z} = (\mathbf{1} \cdot \boldsymbol{z})\mathbf{1}$ with $\boldsymbol{z} = \nabla F(\boldsymbol{v})$.

## Chapter 9

**9.1.**

$$\boldsymbol{x} \cdot Q_0\boldsymbol{x} = (1-\rho)\boldsymbol{x} \cdot \boldsymbol{x} + \rho\boldsymbol{x} \cdot J\boldsymbol{x} = (1-\rho)|\boldsymbol{x}|^2 + \rho(\mathbf{1} \cdot \boldsymbol{x})^2 =$$
$$(1-\rho)(mv(\boldsymbol{x}) + m\bar{x}^2) + \rho m^2\bar{x}^2 = (1-\rho)mv(\boldsymbol{x}) + m(1 + (m-1)\rho)\bar{x}^2.$$

Here we used the identity $J\boldsymbol{x} = (\mathbf{1} \cdot \boldsymbol{x})\mathbf{1}$.

**9.2.** (a) This is a straightforward matrix multiplication.

(b)

$$Q_0^{-1}Q_0 = (1-\rho)^{-1}(I - \kappa J)((1-\rho)I + \rho J) =$$
$$(1-\rho)^{-1}((1-\rho)I + (\rho - \kappa(1-\rho))J - \kappa\rho J^2) =$$
$$(1-\rho)^{-1}((1-\rho)I + (\rho - \kappa(1-\rho))J - \kappa\rho mJ) = I.$$

**9.3.** Let $\boldsymbol{y} = S^{-1}\boldsymbol{x}$. Then $y_i = x_i/\sigma_i$, and

$$Q^{-1}\boldsymbol{x} = S^{-1}Q_0^{-1}\boldsymbol{y} = (1-\rho)^{-1}S^{-1}(\boldsymbol{y} - \kappa J\boldsymbol{y})) = (1-\rho)^{-1}S^{-1}(\boldsymbol{y} - \kappa(\mathbf{1} \cdot \boldsymbol{y})\mathbf{1}),$$

from which the result follows.

**9.4.** We have $v_i = Cw_i$, where $C = \sigma^2_{min}/(1-\rho)$. The sum of the weights of the minimum variance portfolio equals 1. Therefore $C = 1/\sum w_j$, and hence the two identities hold.

**9.5.** We have $R_M = \sum_{i=1}^N p_i R_i$. and hence

$$\sigma_M^2 = \text{Var}(\sum_{i=1}^{N} p_i R_i) = \sum_{i,j} p_i p_j \text{Cov}(R_i, R_j) = \sum_{i=j} p_i p_j \sigma_i \sigma_j + \sum_{i \neq j} p_i p_j \sigma_i \sigma_j \rho =$$

$$\sum_{i=j} p_i p_j \sigma_i \sigma_j - \sum_{i=j} p_i p_j \sigma_i \sigma_j \rho + \sum_{i,j} p_i p_j \sigma_i \sigma_j \rho = (1-\rho)\sum_i (p_i\sigma_i)^2 + \rho(\sum_i p_i \sigma_i)^2.$$

Similarly,

$$\sigma_M^2 \beta_i = \text{Cov}(R_i, \sum_{j=1}^{N} p_j R_j) = \sum_{j=1}^{N} p_j \text{Cov}(R_i, R_j) =$$

$$\sum_{j,j=i} p_j \text{Cov}(R_i, R_j) + \sum_{j,j \neq i} p_j \text{Cov}(R_i, R_j) =$$

$$p_i \sigma_i^2 + \sum_{j,j\neq i} p_j \sigma_i \sigma_j \rho = p_i \sigma_i^2 - p_i \sigma_i^2 \rho + \sum_j p_j \sigma_i \sigma_j \rho = (1-\rho) p_i \sigma_i^2 + \rho \sigma_i \sum_j p_j \sigma_j.$$

**9.6.** The weights of the beta portfolio are proportional to $Q^{-1}\beta$. It therefore follows from (9.4) and (9.8) that $v_\beta(i)$ is proportional to

$$\frac{1}{\sigma_i}(\rho\bar{\sigma} + (1-\rho)p_i\sigma_i - \kappa \sum_{j=1}^{m}(\rho\bar{\sigma} + (1-\rho)p_j\sigma_j)) = (1-\rho)p_i + \frac{K}{\sigma_i}.$$

Here

$$K = \rho\bar{\sigma}(1 - m\kappa) - \kappa(1-\rho)\sum_{j=1}^{m} p_j \sigma_j.$$

The identity

$$\rho(1 - m\kappa) = \kappa(1-\rho)$$

is easily verified, and hence

$$K = (1-\rho)\kappa(\bar{\sigma} - \sum_{j=1}^{m} p_j \sigma_j).$$

The weights are therefore proportional to $(1-\rho)w_i$ and hence also to $w_i$.

**9.7.** Let

$$t = 1/\sigma_1 + \cdots + 1/\sigma_m \text{ and } C = \bar{\sigma}'_m \kappa.$$

Then $C > 0$, and

$$v_\beta(i) = \frac{p_i + \frac{C}{\sigma_i}}{s_m + Ct}.$$

The following inequalities are therefore equivalent

$$v_\beta(i) < p_i/s_m, \; s_m(p_i + C/\sigma_i) < p_i(s_m + Ct),$$

$$s_m C/\sigma_i < p_i Ct, \; 1/(t\sigma_i) < p_i/s_m.$$

**9.8.** Let $I = \{j; (1-\rho)p_j\sigma_j^2/\sigma_M^2 < \epsilon\}$. Then

$$\sum_{j \in I}(1-\rho)p_j^2\sigma_j^2/\sigma_M^2 < \sum_{j \in I} \epsilon p_j \leq \epsilon.$$

**9.9.** Volatility: $0.226$. Weights: $0.30, 0.04, 0.10, 0.05, 0.51$.

**9.10.** If, for example, $m = 3$, $x_1 = 1$, $\frac{\rho}{1+\rho} < x_2 < \rho$, and $x_3 = 0$, then the first $J$ will satisfy $J = \{x_1, x_2\}$ but $I_* = \{x_1\}$.

**9.11.** Equal weights within each group.

| | AZN | ERIC | HM | SDIA | SKA | $\sigma$ |
|---|---|---|---|---|---|---|
| $a$ | 5.7 | −2.8 | −0.9 | −2.6 | 13.1 | 14.6 |
| $b$ | 2.1 | 0 | 0 | 0 | 10.4 | 16.3 |
| $c$ | 2.5 | 2.5 | 2.5 | 2.5 | 2.5 | 21.7 |

The unit is %.

**9.12.** Let $y_i = \sigma_i x_i$.Then

$$\boldsymbol{x} \cdot Q\boldsymbol{x} = \sum_i y_i^2 + \sum_{i \neq j} \rho_i y_i \rho_j y_j = \sum_i (1-\rho_i^2)y_i^2 + (\sum_i \rho_i y_i)^2.$$

(a) Assume that the condition is not satisfied for two values of $i$, $i = 1$, and $i = 2$, say. Let $y_2 = y_3 = \cdots = y_m = 0$, and choose $y_1 \neq 0$ and $y_2 \neq 0$ such that $\rho_1 y_1 + \rho_2 y_2 = 0$. Then

$$\boldsymbol{x} \cdot Q\boldsymbol{x} = (1-\rho_1^2)y_1^2 + (1-\rho_2^2)y_2^2 \leq 0,$$

and hence $Q$ is not strictly positive definite.

(b)

$$\boldsymbol{x} \cdot Q\boldsymbol{x} \geq \sum_i (1-\rho_i^2)y_i^2 > 0,$$

if $\rho_i^2 < 1$ for all $i$ and $\boldsymbol{x} \neq 0$.

**9.13.** (a) This is a straightforward matrix calculation. A first step can be to find that all elements in the $i$th row of $D^2J$ equals $d_i$.

(b)

$$(I - \frac{1}{1+d}DJD)(I + DJD) = I + (1 - \frac{1}{1+d})DJD - \frac{1}{1+d}DJD^2JD =$$

$$I + (1 - \frac{1}{1+d})DJD - \frac{1}{1+d}dDJD = I.$$

Here we used the identity in (a).

**9.14.** We have by Exercise 9.13b that

$$Q^{-1} = \Delta^{-1}(I - \frac{1}{1+d}DJD)\Delta^{-1} = \Delta^{-2} - \frac{1}{1+d}EJE,$$

where $E = D\Delta^{-1}$ is the diagonal matrix with elements

$$e_i = \frac{\rho_i}{1-\rho_i^2}\frac{1}{\sigma_i}.$$

Note that

$$JE\boldsymbol{x} = (\mathbf{1} \cdot E\boldsymbol{x})\mathbf{1},$$

and hence

$$(EJE\boldsymbol{x})_i = (\mathbf{1} \cdot E\boldsymbol{x})(E\mathbf{1})_i = (\sum_{j=1}^m e_j x_j)e_i.$$

Also

$$(\Delta^{-2}\boldsymbol{x})_i = \frac{x_i}{\sigma_i^2(1-\rho_i^2)}.$$

**9.15.** $0.55, 0.54, 0.49, 0.68, 0.41.$

**9.16.** $0.30, 0.04, 0.13, 0, 0.52.$

# REFERENCES


1. Baxter, M., and Rennie A. 1996. *Financial Calculus.* Cambridge University Press, New York.
2. Elton, E. J., Gruber, M. J. 1981 *Modern Portfolio Theory and Investment Analysis.* John Wiley & Sons, New York.
3. Fama, E. F., and French K. R. 2004. The capital asset pricing model: Theory and evidence *Journal of Economic Perspectives* **18**, No. 3, 25–46.
4. Frye, J. 1997. Principals of risk: Finding VAR through factor-based interest rate scenarios. In *VAR: Understanding and Applying Value at Risk.* Risk Publications, London, 275-288.
5. Hull, J. C. 1989. *Options Futures, & Other Derivatives.* Prentice Hall, Englewood Cliffs, NJ.
6. Lang, H. 1999 , Stockholm. Personal communication.
7. Luenberger, D. G. 1998. *Investment Science.* Oxford University Press, New York.
8. Markowitz, H. M. 1952. Portfolio selection. *Journal of Finance*, **7**, 77–91.
9. Markowitz, H. M. 1959. *Portfolio Selection: Efficient Diversification of Investments.* John Wiley & Sons, New York.
10. Perold, A. F. 2004. The capital asset pricing model. *Journal of Economic Perspectives* **18**, No. 3, 3–24.

11. Rogers, L. C. G., and Stapleton E. J. 1998. Fast accurate binomial pricing. *Finance and Stochastics* **2**, 3–17.

# INDEX